Progress in Mathematics
Vol. 49

Edited by
J. Coates and
S. Helgason

Birkhäuser Verlag
Boston · Basel · Stuttgart

Henrik Schlichtkrull

Hyperfunctions and Harmonic Analysis on Symmetric Spaces

1984

Birkhäuser
Boston · Basel · Stuttgart

Author:

Henrik Schlichtkrull
Mathematical Institute
University of Copenhagen
Universitetsparken 5
DK–2100 Copenhagen
Denmark

Library of Congress Cataloging in Publication Data

Schlichtkrull, Henrik, 1954 –
Hyperfunctions and harmonic analysis on symmetric spaces.

(Progress in mathematics ; vol. 49)
Bibliography: p.
Includes Index.
1. Symmetric spaces. 2. Hyperfunctions. 3. Harmonic analysis. I. Titel. II. Series : Progress in mathematics (Boston, Mass.) ; vol. 49
QA649.S34 1984 512'.55 84-11 64
ISBN 0-8176-3215-8

CIP-Kurztitelaufnahme der Deutschen Bibliothek

Schlichtkrull, Henrik:
Hyperfunctions and harmonic analysis on symmetric spaces / Henrik Schlichtkrull. – Boston ; Basel ; Stuttgart : Birkhäuser, 1984.
(Progress in mathematics ; Vol. 49)
ISBN 3-7643-3215-8 (Basel, Stuttgart)
ISBN 0-8176-3215-8 (Boston)
NE: GT

ISBN 0-8176-3215-8
ISBN 3-7643-3215-8
Printed in Switzerland
9 8 7 6 5 4 3 2 1

Acknowledgements

This monograph is an outgrowth of the author's essay "Applications of hyperfunction theory to representations of semisimple Lie groups", written in response to a prize question which was posed by the University of Copenhagen for 1982. I am grateful to the University of Copenhagen for the inspiration it gave me by awarding the essay a gold medal.

The prize essay has been transformed into the present book by an entire rewriting, following numerous improvements suggested by Professors Mogens Flensted-Jensen and Sigurður Helgason, to whom I am profoundly grateful. I am also indebted to Professor Toshio Oshima for permission to give an account of his proof of Theorem 8.3.1, outlined in correspondence with Flensted-Jensen, and for pointing out to me the necessity of using Theorem 2.5.8 in Section 6.3. I am most grateful to Professor Kiyosato Okamoto for helpful introductions to understanding microlocal analysis, and to Professors David Collingwood, Hans Jakobsen and Anthony Knapp who read the manuscript (or parts thereof), pointed out errors, and suggested many improvements. Any error that remains is entirely of my own responsibility.

In the process of writing this book I have been supported by the Danish National Science Research Council, to which I express my gratitude. Most of the work has been done while being a member of the School of Mathematics of The Institute for Advanced Study, whose hospitality I am grateful for. In particular, I am grateful to Ms. Dorothea Phares for her skillful typing of the manuscript.

Finally, I proudly express a profound debt to my wife Birgitte, to whom I dedicate this book.

Henrik Schlichtkrull
Princeton, New Jersey

March, 1984

Til Birgitte

Introduction

The purpose of this book is to give an exposition of the application of hyperfunction theory and microlocal analysis to some important problems in harmonic analysis of symmetric spaces.

The theory of hyperfunctions generalizes that of distributions in the sense that while distributions are linear functionals on C^∞-functions, hyperfunctions can be thought of as linear functionals on the smaller space of analytic functions. For the study of partial differential equations with analytic coefficients this concept is extremely useful. Microlocal analysis is the study (via the tangent space) of the local properties of solutions to systems of such equations.

The book consists of two parts. In the first part (Chapters 1 and 2), which is expository, we give an introduction to hyperfunctions, microlocal analysis, and applications of this theory to the study of systems of partial differential equations with regular singularities. We give very few proofs. As for the main results (Theorems 2.3.1 and 2.3.2), we illustrate the technique of proof via an important example (Section 2.4).

In the second part, we apply the results from the first part to symmetric spaces. Here we give full proofs of all results (with one exception, cf. below); except for certain standard results from the theory of semisimple Lie groups (stated in Chapter 3), this part of the book is self contained (that is, modulo Chapters 1 and 2).

There are two main results that we prove in the second part of the book, concerning respectively a Riemannian symmetric space and a semisimple symmetric space.

Let X be a Riemannian symmetric space of the noncompact type and let $\mathbb{D}(X)$ be the algebra of differential operators on X invariant under all isometries of X. The first result (Corollary 5.4.4) states that every function on X which is an eigenfunction for each operator in $\mathbb{D}(X)$ can be represented by a hyperfunction on the boundary of X via an integral formula similar to the classical Poisson integral for the unit disk. This result, the proof of which comprises Chapters 4 and 5, was conjectured by S. Helgason (1970,[c]) and proved by M. Kashiwara, A. Kowata, K. Minemura, K. Okamoto, T. Oshima and M. Tanaka (1978, Kashiwara et al. [a]) by employing the techniques of microlocal analysis to study the boundary behavior of the eigenfunction to $\mathbb{D}(X)$. This is done by imbedding X into a compact analytic manifold such that the differential operators in $\mathbb{D}(X)$ have regular singularities along the boundary of X (Theorem 4.3.1). The theory from the first part of the book then ensures that the eigenfunctions have certain "boundary values" (or Cauchy data), which are hyperfunctions on the boundary. It is then proved that by taking the Poisson integral of one of these boundary values we recover the original eigenfunction on X (Theorem 5.4.2).

However, in order for the above proof to work, the eigenvalues for the operators in $\mathbb{D}(X)$ have to satisfy a certain regularity assumption (to ensure that no logarithmic terms appear in the process of taking the boundary values). In order to prove Helgason's conjecture for the remaining singular eigenvalues, more refined methods are needed. It is for this most general statement of the conjecture (Theorem 5.4.3) that we make an omission of proof.

In Chapter 6 a generalization of Helgason's conjecture is presented. In the compactification of X (which is known as the maximal Satake-Furstenberg compactification) the so-called boundary of X is in fact only one part of the boundary. The boundary has in general several other "components", and it is natural to represent the eigenfunctions on X also as Poisson integrals of their hyperfunction boundary values on these components (Theorem 6.3.3).

One of the features of the theory of differential equations with regular singularities is that it enables us to derive asymptotic expansions of solutions in the vicinity of the regular singular points. We illustrate this technique by deriving asymptotic expansions of the spherical functions on the Riemannian symmetric space (Theorems 5.3.2 and 6.3.4). These asymptotic expansions (though not in the form of Theorem 6.3.4) were originally derived by Harish-Chandra.

The second main result, concerning a semisimple symmetric space, is proved in Chapters 7 and 8 by using the same technique as was employed in Chapter 6. Let G/H be a semisimple symmetric space (that is, G a semisimple connected Lie group and H a subgroup which is the identity component of the set of fixed points for some involutive autohorphism of G). In the harmonic analysis of G/H one wants to determine the closed subspaces of $L^2(G/H)$ on which G acts irreducibly in the regular representation (that is, the representation of G on $L^2(G/H)$ by left translations) - the so-called discrete series for $L^2(G/H)$. This problem was attacked by M. Flensted-Jensen, who constructed a family of functions on G/H (cf. Section 8.3), which he conjectured to be square integrable (1979, [c]). These functions are eigenfunctions for the invariant differential operators on G/H , and in the "generic" range of the eigenvalue, he proved the square integrability. The conjecture (Theorem 8.3.1) was settled (affirmatively) by T. Oshima (1980, unpublished - cf. Oshima and Matsuki [b]). The proof consists of an application of the theory of regular singularities to derive asymptotic expansions and hence growth estimates for Flensted-Jensen's functions.

The requirements on the part of the reader are as follows. For the hyperfunction theory some familiarity with complex functions of several variables is desirable. However, since this part of the book is expository no deep knowledge is necessary, unless the reader wants to consult the references for proofs. For the applications to symmetric spaces the reader has to be acquainted with some Lie group theory, as for instance is offered in the books Helgason [j] or Wallach [a]. See also Chapter 3 for a more detailed description of the necessary prerequisites.

This book contains several new results. As for the two main results mentioned above, however, the contribution of the author is solely expository. The author's main original contributions are to be found in Chapter 6. Each chapter is concluded with a short section of notes, giving the origin of the theory described in that chapter, with references to the bibliography, which is in the back of the book. The references for the main theorems are Kashiwara and Oshima [a], Oshima and Sekiguchi [a], Oshima [a], Kashiwara et al. [a], Flensted-Jensen [c] and Oshima [c].

Notation.

$\mathbb{R}$ = field of real numbers, $\mathbb{R}_+ = \{t \in \mathbb{R} \mid t \geq 0\}$,

$\mathbb{C}$ = field of complex numbers ,

$\mathbb{Z}$ = ring of integers, $\mathbb{Z}_+ = \mathbb{Z} \cap \mathbb{R}_+$,

$\mathbb{N}$ = set of positive integers.

TABLE OF CONTENTS

1. Hyperfunctions and microlocal analysis - an introduction

The theory of hyperfunctions is in some sense a generalization of the theory of distributions, as developed by L. Schwartz. On a compact real analytic manifold M, a hyperfunction is the same as an analytic functional, that is, a continuous linear functional on the space $\mathcal{A}(M)$ of analytic functions on M (equipped with a certain topology of inductive limits). On noncompact real analytic manifolds hyperfunctions are most conveniently studied by cohomological methods introduced by M. Sato. In the first four sections of this chapter we develop this theory, introducing along the way some elementary sheaf theory. The last two sections consist of a brief introduction to microlocal analysis.

1.1. Hyperfunctions of one variable

The hyperfunctions of one variable are much easier to understand than those of more variables. Therefore, we begin with them.

Let $V \subset \mathbb{R}$ be an open set, and let $W \subset \mathbb{C}$ be a complex neighborhood of V, i.e., W is open and contains V as a closed subset (for example, $V + \sqrt{-1}\,\mathbb{R}$ is a complex neighborhood of V).

By definition, the space $\mathcal{B}(V)$ of hyperfunctions on V is given by

$$(1.1) \qquad \mathcal{B}(V) = \mathcal{O}(W \setminus V)/\mathcal{O}(W)$$

where $\mathcal{O}(U)$ for an open set $U \subset \mathbb{C}$ denotes the space of holomorphic functions on U.

For $F \in \mathcal{O}(W \setminus V)$ we use the notation $[F]$ for the equivalence class of F in $\mathcal{B}(V)$ and call F a defining function for $[F]$.

As already indicated in the notation, we have:

Proposition 1.1.1 The space $\mathcal{B}(V)$ as defined in (1.1) does not depend on the choice of complex neighborhood W.

Proof: Let W' and W be complex neighborhoods of V and assume $W' \subset W$. Induced from restriction, we have a map $\mathcal{O}(W\backslash V)/\mathcal{O}(W) \longrightarrow \mathcal{O}(W'\backslash V)/\mathcal{O}(W')$, which we claim is an isomorphism. The injectivity is clear, and the surjectivity easily follows from the lemma below. □

Lemma 1.1.2 Let $W_1, W_2 \subset \mathbb{C}$ be open sets with $W_1 \cap W_2 \neq \emptyset$. For each function $\varphi \in \mathcal{O}(W_1 \cap W_2)$ there exist functions $\varphi_i \in \mathcal{O}(W_i)$ $(i = 1,2)$ such that $\varphi = \varphi_2 - \varphi_1$ on $W_1 \cap W_2$.

The lemma is a special case of the following theorem, which is a strong form of Mittag-Leffler's theorem, cf. Hörmander [a] Theorem 1.4.5.

Theorem 1.1.3 Let $\{W_\alpha \mid \alpha \in A\}$ be a family of open sets in $\mathbb{C}$, and let $\{\varphi_{\alpha,\beta} \mid \alpha, \beta \in A\}$ be a family of functions $\varphi_{\alpha,\beta} \in \mathcal{O}(W_\alpha \cap W_\beta)$ such that $\varphi_{\alpha,\beta} = -\varphi_{\beta,\alpha}$ and

$$\varphi_{\beta,\gamma} - \varphi_{\alpha,\gamma} + \varphi_{\alpha,\beta} = 0$$

on $W_\alpha \cap W_\beta \cap W_\gamma$ for all $\alpha, \beta, \gamma \in A$. Then there exist holomorphic functions $\varphi_\alpha \in \mathcal{O}(W_\alpha)$ such that

$$\varphi_{\alpha,\beta} = \varphi_\beta - \varphi_\alpha$$

on $W_\alpha \cap W_\beta$ for all $\alpha, \beta \in A$.

Let $V' \subset V$ be an open subset of V, and let $W' = (W\backslash V) \cup V'$. Then W' is a complex neighborhood of V', and $W'\backslash V' = W\backslash V$. We therefore have a canonical mapping from $\mathcal{B}(V) = \mathcal{O}(W\backslash V)/\mathcal{O}(W)$ to $\mathcal{B}(V') = \mathcal{O}(W'\backslash V')/\mathcal{O}(W')$. The map is called restriction and denoted $f \longrightarrow F|_{V'}$. Notice that the restriction is onto $\mathcal{B}(V')$.

Let $V = V_1 \cup V_2$ where V_1 and V_2 are open sets. From the definition of restriction it follows immediately that if $f \in \mathcal{B}(V)$ and $f|_{V_i} = 0$ $(i = 1,2)$, then $f = 0$. Conversely, let $f_i \in \mathcal{B}(V_i)$ $(i = 1,2)$ be given and assume

$$f_1|_{V_1 \cap V_2} = f_2|_{V_1 \cap V_2} .$$

Then we claim that there exists $f \in \mathcal{B}(V)$ such that $f|_{V_i} = f_i$ $(i=1,2)$. To prove this claim, let W_i be a complex neighborhood of V_i such that $V_i = W_i \cap \mathbb{R}$, and let $F_i \in \mathcal{O}(W_i \setminus \mathbb{R})$ be a defining function for f_i $(i=1,2)$. Then, by assumption, there exists $\varphi \in \mathcal{O}(W_1 \cap W_2)$ such that $F_1 - F_2 = \varphi$ on $(W_1 \cap W_2) \setminus \mathbb{R}$. Using Lemma 1.1.2 we write $\varphi = \varphi_2 - \varphi_1$ where $\varphi_i \in \mathcal{O}(W_i)$ $(i=1,2)$. Then $F_1 + \varphi_1 = F_2 + \varphi_2$ on $(W_1 \cap W_2) \setminus \mathbb{R}$, hence we can define $F \in \mathcal{O}((W_1 \cup W_2) \setminus \mathbb{R})$ by $F(Z) = F_i(Z) + \varphi_i(Z)$ for $Z \in W_i \setminus \mathbb{R}$ $(i=1,2)$. With $f = [F]$ we have $f|_{V_i} = f_i$ $(i=1,2)$, as claimed.

The preceding argument is easily generalized to an arbitrary open covering $V = \bigcup_{\alpha \in A} V_\alpha$ of V. The property of hyperfunctions thus obtained is called the <u>localization property</u>.

The <u>support</u> $\operatorname{supp} f$ of a hyperfunction $f \in \mathcal{B}(V)$ is defined as the smallest closed subset C of V such that $f|_{V \setminus C} = 0$. For $C \subset V$ let $\mathcal{B}_C(V) = \{f \in \mathcal{B}(V) \mid \operatorname{supp} f \subset C\}$, and note that if C is closed then $\mathcal{B}_C(V) \simeq \mathcal{B}_C(\mathbb{R})$ because of the localization property.

The most intuitive interpretation of hyperfunctions is as boundary values of holomorphic functions. To describe this, let W be a complex neighborhood of V with $V = W \cap \mathbb{R}$, and let W^+, resp. W^-, be the intersection of W with the upper, resp. lower, half plane. Then if $\varphi \in \mathcal{O}(W^+)$ we define $\Phi \in \mathcal{O}(W \setminus V)$ by $\Phi|_{W^+} = \varphi$ and $\Phi|_{W^-} = 0$. The hyperfunction $[\Phi] \in \mathcal{B}(V)$ is denoted $\varphi(x + \sqrt{-1}\,0)$ and called the <u>boundary value</u> of φ. Similarly, if $\psi \in \mathcal{O}(W^-)$ we define $\Psi \in \mathcal{O}(W \setminus V)$ by $\Psi|_{W^+} = 0$ and $\Psi|_{W^-} = -\psi$ and call $[\Psi] = \psi(x - \sqrt{-1}\,0) \in \mathcal{B}(V)$ the boundary value of ψ.

Let now $f \in \mathcal{B}(V)$ and let $F \in \mathcal{O}(W \setminus V)$ be a defining function for f. If we put $F^+ = F|_{W^+}$ and $F^- = F|_{W^-}$ then we see that

$$f = F^+(x + \sqrt{-1}\,0) - F^-(x - \sqrt{-1}\,0)\,. \tag{1.2}$$

Thus every hyperfunction can be written as the <u>sum of boundary values of holomorphic functions</u> from above and below. Intuitively, f is the <u>jump</u> of F at V.

The space $\mathcal{A}(V)$ of real analytic functions on V can be injected into $\mathcal{B}(V)$ as follows. Let $f \in \mathcal{A}(V)$ and choose the complex neighborhood W of V in such a way that f extends to a holomorphic function F on W. As before, we put $F^+ = F|_{W^+}$ and $F^- = F|_{W^-}$, then since $F \in \mathcal{O}(W)$ we have that $F^+(x + \sqrt{-1}\,0) =$

$F^-(x-\sqrt{-1}\,0)$. The mapping $f \longrightarrow F^+(x+\sqrt{-1}\,0)$ gives the desired injection of $\mathcal{A}(V)$ into $\mathcal{B}(V)$.

Let $f \in \mathcal{A}(V)$ and $g \in \mathcal{B}(V)$. Then we can define the product $fg \in \mathcal{B}(V)$ of f and g as follows: Let W and F be as in the preceding paragraph, let $G \in \mathcal{O}(W\setminus V)$ be a defining function for g , and define $fg = [F\cdot G]$. We differentiate a hyperfunction simply by differentiating the defining function. Thus differential operators with real analytic coefficients operate on hyperfunctions. On $\mathcal{A}(V)$ considered as a subspace of $\mathcal{B}(V)$ these operations have the usual meaning.

Let $f \in \mathcal{B}(V)$ be a hyperfunction and assume that f has support in a compact set K . Let $F \in \mathcal{O}(W\setminus V)$ be a defining function for f , then F extends holomorphically to $W\setminus K$, since $f|_{V\setminus K} = 0$. Assume for simplicity that V and K are intervals, and let γ be a closed curve in $W\setminus K$ encircling K once in the positive direction. Then we define the integral of f by

$$\int f(x)dx = -\int_\gamma F(z)dz \quad .$$

By Cauchy's theorem, the integral does not depend on the choices of either F or γ .

Another consequence of Cauchy's theorem is the following: Let $f \in \mathcal{B}_K(V)$ where $K \subset V$ is compact. For each $z \in \mathbb{C}\setminus V$ define $h_z \in \mathcal{A}(V)$ by

$$h_z(x) = \frac{1}{2\pi\sqrt{-1}(x-z)}$$

and let

$$F(z) = \int (h_z f)(x)dx \quad . \tag{1.3}$$

Then $F \in \mathcal{O}(\mathbb{C}\setminus V)$ and $f = [F]$.

Let $K \subset \mathbb{R}$ be compact and let $\mathcal{A}(K)$ be the space of germs of real analytic functions defined on a neighborhood of K , i.e., $\mathcal{A}(K) = \cup\, \mathcal{A}(V)$, where the union ranges over all open sets $V \subset \mathbb{R}$ containing K and we identify $f_1 \in \mathcal{A}(V_1)$ with $f_2 \in \mathcal{A}(V_2)$ if $f_1 = f_2$ in a neighborhood of K . We endow $\mathcal{A}(V)$ with the topology of uniform convergence on compact sets and $\mathcal{A}(K)$ with the topology of inductive limits. The dual space of $\mathcal{A}(K)$ is denoted $\mathcal{A}'(K)$ and its elements are called analytic functionals on K .

Let $T \in \mathcal{A}'(K)$ be an analytic functional on K. In analogy with (1.3) we define $F(z) = T(h_z)$ for $z \in \mathbb{C} \setminus K$. Then $F \in \mathcal{O}(\mathbb{C} \setminus K)$ so that F defines a hyperfunction $f_T = [F] \in \mathcal{B}_K(\mathbb{R})$.

<u>Proposition 1.1.4</u> <u>The spaces $\mathcal{A}'(K)$ and $\mathcal{B}_K(\mathbb{R})$ are isomorphic via the map $T \longrightarrow f_T$ defined above.</u>

<u>Proof</u>: The following claim immediately shows the map is injective:

(1.4) $$T(\varphi) = \int (\varphi f_T)(x)dx$$

for all $\varphi \in \mathcal{A}(K)$. By definition the right side equals

$$-\int_\gamma \varphi(z) T(h_z) dz$$

where φ has been extended to a holomorphic function on a neighborhood W of K and γ encircles K once in W. By Cauchy's integration formula we have

$$\varphi(x) = -\int_\gamma \varphi(z) h_z(x) dz$$

for $x \in K$. Applying T on both sides and approximating the integral with its Riemann sums it follows as claimed in (1.4) that

$$T(\varphi) = -\int_\gamma \varphi(z) T(h_z) dz \quad .$$

The surjectivity follows from (1.3), using (1.4) to define the inverse. □

Since the space $\mathcal{D}'_K(V)$ of distributions on V with support in K is a subspace of $\mathcal{A}'(K)$ it follows from Proposition 1.1.4 that there is an injection $\mathcal{D}'_K(V) \hookrightarrow \mathcal{B}_K(V)$. Since any distribution can be written as a locally finite sum of distributions with compact support, one can extend this injection to an injection of $\mathcal{D}'(V)$ into $\mathcal{B}(V)$. Thus <u>hyperfunctions generalize distributions</u>.

<u>Examples of hyperfunctions on $\mathbb{R}$</u>. For $z \in \mathbb{C} \setminus \{z < 0\}$ let $\text{Log } z \in \mathbb{C}$ be determined by $z = e^{\text{Log } z}$ and $|\text{ImLog } z| < \pi$.

1) Heaviside's function

$$Y(x) = -\frac{1}{2\pi\sqrt{-1}} [\text{Log}(-z)] \quad .$$

Then

$$Y|_{]-\infty,0[} = 0 \quad \text{and} \quad Y|_{]0,\infty[} = 1$$

2) Dirac's function

$$\delta(x) = -\frac{1}{2\pi\sqrt{-1}}[z^{-1}] .$$

More generally for $n \in \mathbb{Z}_+$

$$\delta^{(n)}(x) = (-1)^{n-1}\frac{n!}{2\pi\sqrt{-1}}[z^{-n-1}] .$$

Then $\delta^{(n)}|_{\mathbb{R}\setminus\{0\}} = 0$, in fact $\delta^{(n)} = \dfrac{d^{n+1}}{dx^{n+1}}Y$.

3) The functions $x_{\pm}^{\lambda}$ $(\lambda \in \mathbb{C}\setminus -\mathbb{N})$ where x_+^{λ} is defined by

$$x_+^{\lambda} = \begin{cases} -\dfrac{1}{2\sqrt{-1}\sin\pi\lambda}[e^{\lambda\,\mathrm{Log}(-z)}] & \text{if } \lambda \in \mathbb{C}\setminus\mathbb{Z} \\ -\dfrac{1}{2\pi\sqrt{-1}}[z^{\lambda}\mathrm{Log}(-z)] & \text{if } \lambda \in \mathbb{Z}_+ \end{cases}$$

and x_-^{λ} by exchanging z with $-z$ in these formulas.

4) The function $e^{1/x} = [e^{1/z}]$ is an example of a hyperfunction which is not a distribution.

Consider another one-dimensional real analytic manifold, the torus $T = \{z \in \mathbb{C} \mid |z| = 1\}$. In analogy with (1.1), we define the space of hyperfunctions on T by

$$\mathcal{B}(T) = \mathcal{O}(\mathbb{C}\setminus T)/\mathcal{O}(\mathbb{C}) .$$

For this space, one can develop a theory analogous to what was done above for $\mathcal{B}(V)$, $V \subset \mathbb{R}$. For instance, one gets that $\mathcal{B}(T) \cong \mathcal{A}'(T)$ similarly to Proposition 1.1.4. One can also use power series to describe $\mathcal{B}(T)$, since the space $\mathcal{O}(\mathbb{C}\setminus T)$ is easily described by power series. In this way, $\mathcal{B}(T)$ is identified with the space of all Fourier series

$$\Sigma_{n=-\infty}^{\infty}\alpha_n e^{in\theta}$$

satisfying the growth condition

$$\forall r>1 \ \exists C>0 \ \forall n \in \mathbb{Z} : |\alpha_n| \le C\cdot r^{|n|} .$$

(Comparing with Helgason [m] Lemma 4.25 this also proves that $\mathcal{B}(T) \cong \mathcal{A}'(T)$) .

We can also describe the space $\mathcal{B}(T)$ as follows: Let $D = \{z \in \mathbb{C} \mid |z| < 1\}$ be the disk and let $\mathcal{H}(D)$ denote the space of harmonic functions on D, that is, all functions annihilated by $\partial^2/\partial x^2 + \partial^2/\partial y^2$ where $z = x + \sqrt{-1}\, y$. Then each $F \in \mathcal{H}(D)$ is real analytic in x and y, and it is easily seen that F has a power series expansion $F(z) = \Sigma_{-\infty}^{\infty} \alpha_n z^n$ satisfying the same growth condition as above. Thus $\mathcal{B}(T) \simeq \mathcal{H}(D)$. The isomorphism is given by the classical Poisson integral (cf. Rudin [a], Chapter 11)

$$\mathcal{P}f(z) = \int_T f(t) \frac{1-|z|^2}{|t-z|^2}\, dt .$$

The integral makes sense for $f \in \mathcal{B}(T)$, because its kernel is analytic. One of our main objects in this book is to generalize this isomorphism (Theorem 5.4.3).

1.2. Sheaves

As mentioned, hyperfunctions in dimension higher than one are more complicated than in dimension one. This is not surprising, since also analytic function theory is more complicated for several variables. The basic tool for handling these difficulties is the theory of sheaves and their cohomology groups. In this section we give the basic definitions from sheaf theory, and in the next section we discuss cohomology of sheaves.

Let X be a topological space.

A sheaf (of abelian groups) on X is a pair $(\mathcal{S}, \pi)$, consisting of a topological space $\mathcal{S}$ and a mapping $\pi : \mathcal{S} \longrightarrow X$ of $\mathcal{S}$ onto X, such that the following (i)-(iii) hold.

(i) π is a local homeomorphism, that is, for each $p \in \mathcal{S}$ there exists a neighborhood Ω of p such that $\pi : \Omega \longrightarrow \pi(\Omega)$ is a homeomorphism.

(ii) Each set $\mathcal{S}_x = \pi^{-1}(x)$, for $x \in X$, is an abelian group (additively written and called the stalk of $\mathcal{S}$ at x).

(iii) The group actions are continuous, that is, the map $(p,q) \longrightarrow p-q$ is continuous from the set $\{(p,q) \in \mathcal{S} \times \mathcal{S} \mid \pi(p) = \pi(q)\}$ (with the relative topology in $\mathcal{S} \times \mathcal{S}$) to $\mathcal{S}$.

If $\mathcal{R}$ and $\mathcal{S}$ are sheaves on X, a morphism of sheaves is a continuous map Φ from $\mathcal{R}$ to $\mathcal{S}$ such that Φ maps the stalk $\mathcal{R}_x$ by a homomorphism into the stalk $\mathcal{S}_x$ for each $x \in X$.

Thus we obtain a category $\mathcal{G}$, the category of sheaves on X.

In most applications, the stalks of a sheaf carry more algebraic structure than just that of abelian groups. It is then required in analogy with (iii) that the algebraic operations are continuous. In this fashion one obtains notions such as those of a sheaf of complex vector spaces, a sheaf of rings, etc. Also, if $\mathcal{R}$ is a sheaf of rings on X, then a sheaf of $\mathcal{R}$-modules is a sheaf $\mathcal{S}$ on X such that each stalk $\mathcal{S}_x$ has the structure of an $\mathcal{R}_x$-module and $(r,p) \longrightarrow rp$ is continuous from $\{(r,p) \in \mathcal{R} \times \mathcal{S} \mid \pi_{\mathcal{R}}(r) = \pi_{\mathcal{S}}(p)\}$ to $\mathcal{S}$.

A presheaf (of abelian groups) F on X is an assignment for each open set $U \subset X$ of an abelian group $F(U)$, and for each pair of open sets $V \subset U$ a homomorphism

$$\rho_V^U : F(U) \longrightarrow F(V)$$

satisfying $F(\emptyset) = \{0\}$, $\rho_U^U = \mathrm{id}$ and $\rho_W^V \circ \rho_V^U = \rho_W^U$ whenever $W \subset V \subset U$.

We call ρ_V^U the restriction map and often denote, for $f \in F(U)$, the image $\rho_V^U(f)$ by $f|_V$.

If F and G are presheaves on X, a morphism of presheaves from F to G is an assignment to each open set $U \subset X$ of a homomorphism $\Phi_U : F(U) \longrightarrow G(U)$ such that $\Phi_V \circ \rho_V^U = \rho_V^U \circ \Phi_U$ whenever $V \subset U$.

Thus we obtain a category $\mathfrak{F}$, the category of presheaves on X.

Again, additional algebraic structure can be added to the groups $F(U)$ with the requirement that the restriction maps preserve this structure, to obtain presheaves of vector spaces, etc.

Let $\mathcal{S}$ be a sheaf on X and $U \subset X$ an open set. A continuous map $s : U \longrightarrow \mathcal{S}$ is called a section over U if $\pi \circ s$ is the identity on U. The set of sections over U is denoted $\Gamma(U,\mathcal{S})$ or just $\mathcal{S}(U)$. We define $\Gamma(\emptyset,\mathcal{S}) = \{0\}$. Since there is an obvious map of restrictions from $\Gamma(U,\mathcal{S})$ to $\Gamma(V,\mathcal{S})$ for $V \subset U$, the assignment of $\Gamma(U,\mathcal{S})$ to U gives a presheaf $\Gamma(\mathcal{S})$ on X, the presheaf of sections of $\mathcal{S}$. Γ is a covariant functor from the category $\mathcal{G}$ to the category $\mathcal{F}$.

Conversely, let F be a presheaf. Then we can construct a sheaf as follows:

For every $x \in X$ let F_x be the direct limit of the spaces $F(U)$ where U ranges over the open neighborhoods of x . That is, in the disjoint union of all $F(U)$ with $x \in U$ an equivalence relation is defined by $f_1 \sim f_2$ if $f_i \in F(U_i)$ $(i = 1,2)$ and $f_1|_V = f_2|_V$ for some neighborhood V of x contained in $U_1 \cap U_2$. Then $F_x = \cup_{x \in U} F(U)/\sim$. The equivalence class of $f \in F(U)$ in F_x is called the <u>germ</u> of f at x and is denoted f_x .

Now let $\mathcal{S} = \bigcup_{x \in U} F_x$ (disjoint union) and define $\pi : \mathcal{S} \longrightarrow X$ by $\pi(F_x) = \{x\}$. On $\mathcal{S}$ we introduce the topology generated by all sets

$$O(U,f) = \{f_x \mid x \in U\}$$

where U ranges over all open sets in X and f over $F(U)$. Then it is easy to verify that $(\mathcal{S},\pi)$ is a sheaf. It is called the <u>sheaf of germs</u> associated to F . If we put $\Gamma^\vee(F) = \mathcal{S}$, then $\Gamma^\vee$ is a covariant functor from $\mathcal{F}$ to $\mathcal{G}$.

Notice that in the preceding construction, for each open set $U \subset X$ there is a natural map $f \longrightarrow s$ from $F(U)$ to $\Gamma(U, \mathcal{S})$ given by $s(x) = f_x$. It is easy to see that $f \longrightarrow s$ is an isomorphism for each U if and only if the presheaf F satisfies the following <u>localization property</u>:

For each family $(U_i)_{i \in I}$ of open sets in X and each family $(f_i)_{i \in I}$ of elements $f_i \in F(U_i)$ such that $f_i|_{U_i \cap U_j} = f_j|_{U_i \cap U_j}$ for all $i,j \in I$, there exists a unique element $s \in F(\bigcup_{i \in I} U_i)$ with $f|_{U_i} = f_i$ for all $i \in I$.

If this property holds, we identify sections of $\mathcal{S}$ over U with elements of $F(U)$ via the natural isomorphism. In other words, $\Gamma(U,\Gamma^\vee(F)) = F(U)$ if F satisfies the localization property.

Conversely, for every sheaf $\mathcal{S}$, $\Gamma^\vee(\Gamma(\mathcal{S}))$ is naturally isomorphic to $\mathcal{S}$, as easily follows from the construction.

<u>Example 1.2.1</u> Let $X = \mathbb{C}^n$ and for each open set $U \subset X$ let $F(U)$ consist of the holomorphic functions on U . Then, with the obvious map of restrictions, F is a presheaf, and it has the

localization property. The associated sheaf $\Gamma^{\vee}(F)$ is denoted $\mathcal{O}$ and called the sheaf of germs of holomorphic functions. The set $\mathcal{O}(U)$ of sections over U is identical to $F(U)$, because of the localization property.

Example 1.2.2 Let $X = \mathbb{R}$, $U \subset X$ open, and let $F(U)$ consist of the bounded functions on U. Then F is a presheaf, but it does not have the localization property. The space of sections over U of the associated sheaf of germs can be identified with the space of locally bounded functions on U.

Example 1.2.3 Let $X = \mathbb{R}$ and let $F(U)$ consist of the hyperfunctions on U. Then F is a presheaf, and as already indicated in Section 1.1, F has the localization property. The associated sheaf $\mathcal{B}$ is called the sheaf(of germs) of hyperfunctions on $\mathbb{R}$. The space $\mathcal{B}(U)$ of its sections is precisely the space of hyperfunctions on U.

If $(\mathcal{S},\pi)$ is a sheaf on X and $\mathcal{R} \subset \mathcal{S}$ is a subset, we call $\mathcal{R}$ a subsheaf of $\mathcal{S}$ if $\mathcal{R}_x = \mathcal{S}_x \cap \mathcal{R}$ is a subgroup of $\mathcal{S}_x$ for each $x \in X$, and moreover $(\mathcal{R}, \pi|_{\mathcal{R}})$ is a sheaf.

If $\mathcal{R}$ is a subsheaf of $\mathcal{S}$, let $\mathcal{S}/\mathcal{R} = \bigcup_{x \in X} \mathcal{S}_x/\mathcal{R}_x$ (disjoint union) and define $p : \mathcal{S} \longrightarrow \mathcal{S}/\mathcal{R}$ stalkwise via the projection $\mathcal{S}_x \longrightarrow \mathcal{S}_x/\mathcal{R}_x$. Equip $\mathcal{S}/\mathcal{R}$ with the topology for which $W \subset \mathcal{S}/\mathcal{R}$ is open if and only if $p^{-1}(W) \subset \mathcal{S}$ is open. Then $\mathcal{S}/\mathcal{R}$ is a sheaf, called the quotient sheaf.

If $\mathcal{R}$ and $\mathcal{S}$ are sheaves on X and $\Phi : \mathcal{R} \longrightarrow \mathcal{S}$ is a morphism, then the kernel, $\ker \Phi = \{r \in \mathcal{R} \mid \Phi(r) = 0\}$, and the image, $\operatorname{im} \Phi = \Phi(\mathcal{R})$, are subsheaves of $\mathcal{R}$ and $\mathcal{S}$ respectively.

Similarly, one can define notions for presheaves of subpresheaf, quotient presheaf, etc.

Recall that a covariant functor F from one category to another is called left (resp. right) exact if for each exact sequence $0 \to A \to B \to C$ (resp. $A \to B \to C \to 0$) the induced sequence $0 \to FA \to FB \to FC$ (resp. $FA \to FB \to FC \to 0$) is exact. Also, F is called exact if it is both right and left exact.

The proof of the following lemma is easy:

Lemma 1.2.4 The functor Γ from $\mathfrak{G}$ to $\mathfrak{F}$ is left exact. The functor $\Gamma^{\vee}$ from $\mathfrak{F}$ to $\mathfrak{G}$ is exact.

It is of crucial importance for the following section to notice that Γ is not right exact in general. This means that, if $\mathcal{R}$ is a subsheaf of $\mathcal{A}$, then the quotient presheaf $\Gamma(\mathcal{A})/\Gamma(\mathcal{R})$ does not satisfy the localization property and thus $\Gamma(\mathcal{A}/\mathcal{R})$ is different from $\Gamma(\mathcal{A})/\Gamma(\mathcal{R})$, in general.

A sheaf $\mathcal{A}$ on X is called flabby if, for every open set $U \subset X$, the restriction $\mathcal{A}(X) \longrightarrow \mathcal{A}(U)$ is surjective.

From Section 1.1 we see that the sheaf $\mathcal{B}$ of hyperfunctions on $\mathbb{R}$ is flabby, whereas, obviously, the sheaf $\mathcal{D}'$ of distributions on $\mathbb{R}$ is not flabby.

Let $\mathcal{A}$ be a sheaf on X. There is a flabby sheaf $\mathcal{A}_0$ containing $\mathcal{A}$ as a subsheaf, which is constructed as follows: Let F be the presheaf for which $F(U)$ consists of all maps f from U to $\mathcal{A}$ such that $\pi \circ f = \mathrm{id}$. Then $\mathcal{A}_0$ is defined by $\mathcal{A}_0 = \Gamma^{\vee}(F)$. It is easily seen that F has the localization property, that $\mathcal{A}_0$ is flabby and that there is a canonical injection of $\mathcal{A}$ into $\mathcal{A}_0$. Also, $\mathcal{A}_0$ has the following property: If $\mathcal{R}$ is a sheaf and $\Phi : \mathcal{R} \longrightarrow \mathcal{A}_0$ is a map such that $\Phi_x = \Phi|_{\mathcal{R}_x}$ is a homomorphism into $\mathcal{A}_{0x}$ for each $x \in X$, then Φ is a morphism of sheaves.

1.3. Cohomology of sheaves

Let $\mathfrak{G}$ denote the category of sheaves of complex vector spaces on X. For $\mathcal{A} \in \mathfrak{G}$ we have constructed a flabby sheaf $\mathcal{A}_0$ in the preceding section.

Lemma 1.3.1 $\mathcal{A}_0$ is injective in $\mathfrak{G}$ for every $\mathcal{A} \in \mathfrak{G}$, that is, for every sheaf $\mathcal{R}$ and subsheaf $\mathcal{R}'$ each morphism $\Phi : \mathcal{R}' \longrightarrow \mathcal{A}_0$ extends to a morphism $\mathcal{R} \longrightarrow \mathcal{A}_0$.

Proof: We extend Φ stalkwise to a map $\Phi^{\sim} : \mathcal{R} \longrightarrow \mathcal{A}_0$ which is linear on each stalk. Then $\Phi^{\sim}$ is a morphism by the final remark of Section 1.2. $\square$

Thus the category $\mathfrak{G}$ has "enough injectives" in the sense that every object $\mathcal{A}$ of $\mathfrak{G}$ can be imbedded into an injective one.

Let $\mathcal{S} \in \mathfrak{S}$, then we have the exact sequence $0 \to \mathcal{S} \to \mathcal{S}_0$. Let $\mathcal{R}$ be the quotient sheaf $\mathcal{R} = \mathcal{S}_0/\mathcal{S}$ and let $\mathcal{S}_1 = \mathcal{R}_0$, then we get an exact sequence $0 \to \mathcal{S} \to \mathcal{S}_0 \to \mathcal{S}_1$. Iterating this procedure we have an exact sequence

$$0 \to \mathcal{S} \to \mathcal{S}_0 \to \mathcal{S}_1 \to \mathcal{S}_2 \to \cdots$$

of flabby sheaves $\mathcal{S}_0, \mathcal{S}_1 \cdots$, injective in category $\mathfrak{S}$. This complex is called the flabby resolution of $\mathcal{S}$.

As we have seen, the functor Γ from $\mathfrak{S}$ to $\mathfrak{F}$ is only left exact. Since $\mathfrak{S}$ has enough injectives it therefore makes sense to form the right derived functors $R^n\Gamma (n = 0,1,\ldots)$. These are functors from $\mathfrak{S}$ to $\mathfrak{F}$, and they can be defined as follows: If $\mathcal{S} \in \mathfrak{S}$ then the complex, derived by applying Γ to the flabby resolution,

$$0 \overset{\delta^{-1}}{\to} \Gamma\mathcal{S}_0 \overset{\delta^0}{\to} \Gamma\mathcal{S}_1 \overset{\delta^1}{\to} \Gamma\mathcal{S}_2 \to \cdots$$

is in general not exact since Γ is not right exact. The presheaf $R^n\Gamma\mathcal{S}$ is the n'th homology of this complex:

$$R^n\Gamma\mathcal{S} = \ker \delta^n / \operatorname{im} \delta^{n-1} \quad (n = 0,1,\ldots) .$$

We call $H^n(U;\mathcal{S}) = R^n\Gamma\mathcal{S}(U)$ for $U \subset X$ open the n'th cohomology space of $\mathcal{S}$ on U. The left exactness of Γ implies $H^0(U;\mathcal{S}) = \mathcal{S}(U)$. The n'th cohomology sheaf is the sheaf of germs associated to $R^n\Gamma\mathcal{S}$, this is denoted $\mathcal{H}^n(\mathcal{S})$.

In fact, we need a more general cohomology theory, called local cohomology, but before going on with that we will discuss a different and probably more intuitive way of constructing cohomology of sheaves, the Čech-cohomology theory.

Let $\mathcal{W} = \{W_i\}_{i \in I}$ be an open covering of X. For each non-negative integer p an alternating p-cochain φ is a map which assigns to each $(p+1)$-tuple $(i_0,\ldots,i_p) \in I^{p+1}$ a section $\varphi_{i_0,\ldots,i_p} \in \mathcal{S}(W_{i_0} \cap \ldots \cap W_{i_p})$ such that φ changes sign if two indices are permuted. By $C^p(\mathcal{W};\mathcal{S})$ we denote the space of all alternating p-cochains.

We define the coboundary operator d^p from $C^p(\mathcal{W};\mathcal{S})$ to $C^{p+1}(\mathcal{W};\mathcal{S})$ by

$$(d^p\varphi)_{i_0,\dots,i_{p+1}} = \Sigma_{j=0}^{p+1}(-1)^j\, \varphi_{i_0,\dots,\hat{i_j},\dots,i_{p+1}}|W_{i_0}\cap\dots\cap W_{i_{p+1}} ,$$

where the notation $\hat{i_j}$ indicates that this index shall be removed. From the definition of d it follows easily that $d^p \circ d^{p-1} = 0$, and thus we can define the n'th Čech-cohomology space of $\mathcal{W}$ with coefficients in $\mathcal{S}$:

$$\check{H}^n(\mathcal{W};\mathcal{S}) = \ker d^n/\operatorname{im} d^{n-1} .$$

Suppose now that $\mathcal{V} = \{V_j\}_{j\in J}$ is another open covering of X which is finer than $\mathcal{W}$ (i.e., each V_j is contained in some W_i). It is not difficult to see that this implies that there is a canonical linear map

$$h^n_{\mathcal{W},\mathcal{V}} : \check{H}^n(\mathcal{W};\mathcal{S}) \longrightarrow \check{H}^n(\mathcal{V};\mathcal{S})$$

(cf. Hörmander [a] Prop. 7.3.1). Then the system $\{H^n(\mathcal{W};\mathcal{S}), h^n_{\mathcal{W},\mathcal{V}}\}$ is directed under the ordering "$\mathcal{V}$ finer than $\mathcal{W}$" and we can take the inductive limit (cf. loc. cit. p. 174)

$$\check{H}^n(X;\mathcal{S}) = \underset{\longrightarrow}{\lim}\; H^n(\mathcal{W};\mathcal{S}) .$$

This vector space is called the n'th Čech-cohomology space of $\mathcal{S}$ on X .

Example 1.3.2 Let X be an open set in $\mathbb{C}$. It is easily seen that Theorem 1.1.3 is equivalent to

$$\check{H}^1(\mathcal{W};\mathcal{O}) = 0$$

for all open coverings $\mathcal{W}$ of X . Therefore $\check{H}^1(X;\mathcal{O}) = 0$. More generally, one can prove that $\check{H}^p(X;\mathcal{O}) = 0$ for all $p \geq 1$ (cf. Hörmander [a] Cor. 7.4.2).

Example 1.3.3 Let X be an open set in $\mathbb{C}^n$ where $n > 1$. We have, in analogy with the preceding example, that $\check{H}^p(X;\mathcal{O}) = 0$ for all $p \geq n$ (Malgrange [a] or Schapira[a] p. 120). It is not true for general X that $\check{H}^1(X;\mathcal{O})$ vanishes. However, if X is a domain of holomorphy then $\check{H}^p(X;\mathcal{O}) = 0$ for all $p \geq 1$ (Hörmander [a], Cor. 7.4.2).

The covering $\mathcal{W}$ is called acyclic (for $\mathcal{S}$) if $\check{H}^n(W_{i_0}\cap\dots\cap W_{i_p};\mathcal{S}) = 0$ for all $i_0,\dots,i_p \in I$ and all $p > 0$.

Theorem 1.3.4 (Leray's theorem) Assume that X is paracompact. Then $\check{H}^n(X;\mathcal{S})$ and $H^n(X;\mathcal{S})$ are naturally isomorphic for all $n \geq 0$ and all sheaves $\mathcal{S}$ on X. Moreover, if $\mathcal{U}$ is an acyclic covering of X then $\check{H}^n(\mathcal{U};\mathcal{S})$ is naturally isomorphic to $\check{H}^n(X;\mathcal{S})$.

Proof: See Grauert and Remmert [a] p. 43-44. □

Example 1.3.5 Let $X \subset \mathbb{C}^n$ be open, and let $\mathcal{U}$ be a covering of X with domains of holomorphy (for instance with open balls). Then $\mathcal{U}$ is acyclic for $\mathcal{O}$ by Example 1.3.3 and thus

$$H^n(X;\mathcal{O}) = \check{H}^n(\mathcal{U};\mathcal{O}) .$$

Now let Y be a locally closed subset of X, that is, Y is the intersection of a closed and an open subset of X. For each open subset V of Y (with the relative topology) there exists an open set U in X which contains V as a closed subset. The space of all sections of $\mathcal{S}$ on U whose support is contained in V is easily seen to be independent of the choice of U, and we denote this space by $\Gamma_Y\mathcal{S}(V)$ (although it really does not depend on Y).

Then Γ_Y gives a left exact functor from $\mathfrak{S}$ to presheaves on Y. Again we form the right derived functors $R^n\Gamma_Y$ $(n = 0,1,\dots)$. For $V \subset Y$ open we denote by $H^n_V(X;\mathcal{S})$ the space $R^n\Gamma_Y\mathcal{S}(V)$ and call this the n'th local cohomology space of $\mathcal{S}$ on V. It is given by the n'th homology of the complex:

$$0 \to \Gamma_Y\mathcal{S}_0(V) \to \Gamma_Y\mathcal{S}_1(V) \to \Gamma_Y\mathcal{S}_2(V) \to \dots$$

(and is thus in fact independent of Y). In particular, $H^0_V(X;\mathcal{S}) = \Gamma_Y\mathcal{S}(V)$. The sheaf on Y corresponding to the presheaf $V \to H^n_V(X;\mathcal{S})$ is denoted $\mathcal{H}^n_Y(\mathcal{S})$. If we take $Y = X$ we regain the ordinary cohomology theory.

The following result will be of importance in Section 1.4:

Theorem 1.3.6 Let $Y \subset X$ be locally closed and assume for some n that the local cohomology sheaves $\mathcal{H}^p_Y(\mathcal{S})$ vanish for all $p < n$. Then the presheaves $H^p_V(X;\mathcal{S})$ vanish for $p < n$, and the presheaf $H^n_V(X;\mathcal{S})$ satisfies the localization property. Thus $H^n_V(X;\mathcal{S})$ is the space of sections over $V \subset Y$ of $\mathcal{H}^n_Y(\mathcal{S})$.

Proof: See Schapira [a] p. 34 . □

We will now discuss the analogue of Čech-cohomology for local cohomology. Let Y be a closed subspace of X. By a relative covering of X (modulo Y) we will mean a tuple $(\mathcal{W}, \mathcal{W}')$ consisting of an open covering $\mathcal{W}' = \{W_i\}_{i \in I'}$ of $X \setminus Y$ and an open covering $\mathcal{W} = \{W_i\}_{i \in I}$ of X containing $\mathcal{W}'$, i.e., such that $I' \subset I$. Let $(\mathcal{W}, \mathcal{W}')$ be a relative covering. A relative alternating p-cochain is then an element φ of $C^p(\mathcal{W}; \mathcal{S})$ satisfying $\varphi_{i_0, \ldots, i_p} = 0$ whenever $i_0, \ldots, i_p \in I'$. The subspace of $C^p(\mathcal{W}; \mathcal{S})$ consisting of these we denote by $C^p(\mathcal{W}, \mathcal{W}'; \mathcal{S})$. It is easily seen that the coboundary operator d^p maps $C^p(\mathcal{W}, \mathcal{W}'; \mathcal{S})$ into $C^{p+1}(\mathcal{W}, \mathcal{W}'; \mathcal{S})$ and thus we get a complex

$$0 \to C^0(\mathcal{W}, \mathcal{W}'; \mathcal{S}) \to C^1(\mathcal{W}, \mathcal{W}'; \mathcal{S}) \to \ldots$$

We define $\check{H}^n(\mathcal{W}, \mathcal{W}'; \mathcal{S})$ to be the n'th homology space of this complex.

If $(\mathcal{V}, \mathcal{V}')$ is another relative covering such that $\mathcal{V}$ and $\mathcal{V}'$ are finer than $\mathcal{W}$ and $\mathcal{W}'$ respectively, then we get a homomorphism from $\check{H}^n(\mathcal{W}, \mathcal{W}'; \mathcal{S})$ to $\check{H}^n(\mathcal{V}, \mathcal{V}'; \mathcal{S})$ and therefore we can define the n'th local Čech-cohomology space of $\mathcal{S}$ on Y as the inductive limit:

$$\check{H}^n_Y(X; \mathcal{S}) = \varinjlim \check{H}^n(\mathcal{W}, \mathcal{W}'; \mathcal{S}) .$$

In analogy with Theorem 1.3.4 we have the following. We assume that X and all its open subsets are paracompact.

Theorem 1.3.7 Let $Y \subset X$ be closed. Then $\check{H}^n_Y(X; \mathcal{S})$ and $H^n_Y(X; \mathcal{S})$ are naturally isomorphic for all $n \geq 0$ and all sheaves $\mathcal{S}$ on X. Moreover, if $(\mathcal{W}, \mathcal{W}')$ is a relative covering of X modulo Y such that $\mathcal{W}$ is acyclic then $\check{H}^n(\mathcal{W}, \mathcal{W}'; \mathcal{S})$ is naturally isomorphic to $\check{H}^n_Y(X; \mathcal{S})$.

Proof: Komatsu [d], Theorem 1.10. □

Example 1.3.8 Let X be an open subset of $\mathbb{C}^n$, and let $Y \subset X$ be closed. We get a relative covering by taking

$\mathcal{W}' = \{X \setminus Y\}$ and $\mathcal{W} = \{X\} \cup \mathcal{W}'$. It is then easily seen that $\check{H}^p(\mathcal{W},\mathcal{W}';\mathcal{O}) = 0$ if $p \neq 1$ and $\check{H}^1(\mathcal{W},\mathcal{W}';\mathcal{O}) \cong \mathcal{O}(X \setminus Y)/\mathcal{O}(X)$. If we assume that both $X \setminus Y$ and X are domains of holomorphy it follows that

$$H^p_Y(X;\mathcal{O}) = \begin{cases} \mathcal{O}(X \setminus Y)/\mathcal{O}(X) & \text{if } p = 1 \\ 0 & \text{otherwise.} \end{cases}$$

Example 1.3.9 Let $X = \mathbb{C}^2$ and let $Y = \{(z_1,0) \in X \mid z_1 \in \mathbb{C}\}$. For each open subset V of Y let $U \subset X$ be the set of points (z_1,z_2) such that $(z_1,0) \in V$. Then both $U \cong V \times \mathbb{C}$ and $U \setminus V \cong V \times \mathbb{C} \setminus \{0\}$ are domains of holomorphy and from the preceding example we get that

$$H^p_V(\mathbb{C}^2;\mathcal{O}) = \begin{cases} \mathcal{O}(U \setminus V)/\mathcal{O}(U) & \text{if } p = 1 \\ 0 & \text{otherwise.} \end{cases}$$

From Theorem 1.3.6 it follows that $\mathcal{O}(U \setminus V)/\mathcal{O}(U)$ is the space of sections over V of the sheaf $\mathcal{H}^1_Y(\mathcal{O})$.

Example 1.3.10 Let $X = \mathbb{C}$ and let $Y = \mathbb{R}$. Let $V \subset Y$ be an open set and let W be a complex neighborhood. From Example 1.3.8 we get

$$H^p_V(\mathbb{C};\mathcal{O}) = \begin{cases} \mathcal{O}(W \setminus V)/\mathcal{O}(W) & \text{if } p = 1 \\ 0 & \text{otherwise,} \end{cases}$$

and by Theorem 1.3.6 $\mathcal{O}(W \setminus V)/\mathcal{O}(W)$ is the space of sections over V of the sheaf $\mathcal{H}^1_{\mathbb{R}}(\mathcal{O})$. Thus we have the identity of sheaves on $\mathbb{R}$:

$$\mathcal{B} = \mathcal{H}^1_{\mathbb{R}}(\mathcal{O}) \quad .$$

Example 1.3.11 Let $K \subset \mathbb{R}^n$ be a compact set. The space $\mathcal{A}'(K)$ of analytic functionals on K is defined as follows: Let $\mathcal{A}(K)$ denote the space of germs of real analytic functions on neighborhoods of K endowed with the topology of inductive limits, then $\mathcal{A}'(K)$ is the dual space. This space has a simple description in terms of local cohomology, due to Martineau [a]:

$$H^p_K(\mathbb{C}^n;\mathcal{O}) = \begin{cases} \mathcal{A}'(K) & \text{if } p = n \\ 0 & \text{otherwise.} \end{cases}$$

(see Schapira [a] p. 121).

1.4. Hyperfunctions of several variables

In Example 1.3.10 the definition of hyperfunctions on $\mathbb{R}$ was given an interpretation in terms of local cohomology. In the case of several variables, hyperfunctions are defined in analogy with this interpretation.

Definition. The sheaf $\mathcal{B}$ of hyperfunctions on $\mathbb{R}^n$ is the n'th local cohomology sheaf on $\mathbb{R}^n$ of the sheaf $\mathcal{O}$ of germs of holomorphic functions on $\mathbb{C}^n$:

$$\mathcal{B} = \mathcal{H}^n_{\mathbb{R}^n}(\mathcal{O}) .$$

Thus $\mathcal{B}$ is the sheaf on $\mathbb{R}^n$ associated with the presheaf $V \to H^n_V(\mathbb{C}^n;\mathcal{O})$. Let $V \subset \mathbb{R}^n$ be an open set. For a closed subset S of V we denote by $\mathcal{B}_S(V)$ the space of hyperfunctions on V whose support is contained in S .

Theorem 1.4.1 (i) $H^p_V(\mathbb{C}^n;\mathcal{O}) = 0$ if $p \neq n$.

(ii) $\mathcal{B}(V) = H^n_V(\mathbb{C}^n;\mathcal{O})$.

(iii) If V is bounded then $\mathcal{B}(V) \cong \mathcal{A}'(\bar{V})/\mathcal{A}'(\partial V)$.

(iv) If K is compact in V then $\mathcal{B}_K(V) \cong \mathcal{A}'(K)$.

(v) The sheaf $\mathcal{B}$ is flabby .

Proof: Assume V is bounded. Along with the functors $\Gamma_{\partial V}$, $\Gamma_{\bar{V}}$ and Γ_V there is a long exact sequence (cf. Hilton and Stammbach [a] Theorem IV 6.3)

$$0 \to H^0_{\partial V}(\mathbb{C}^n;\mathcal{O}) \to H^0_{\bar{V}}(\mathbb{C}^n;\mathcal{O}) \to H^0_V(\mathbb{C}^n;\mathcal{O}) \to H^1_{\partial V}(\mathbb{C}^n;\mathcal{O}) \to \dots$$

From Example 1.3.11 we have that $H^p_{\partial V}(\mathbb{C}^n;\mathcal{O}) = H^p_{\bar{V}}(\mathbb{C}^n;\mathcal{O}) = 0$ if $p \neq n$ and $H^n_{\partial V}(\mathbb{C}^n;\mathcal{O}) \cong \mathcal{A}'(\partial V)$ and $H^n_{\bar{V}}(\mathbb{C}^n;\mathcal{O}) \cong \mathcal{A}'(\bar{V})$. Inserting this into the sequence we get that $H^p_V(\mathbb{C}^n;\mathcal{O}) = 0$ if $p \neq n-1,n$ and moreover the exact sequence:

$$0 \to H_V^{n-1}(\mathbb{C}^n;\mathcal{O}) \to \mathcal{A}'(\partial V) \to \mathcal{A}'(\bar{V}) \to H_V^n(\mathbb{C}^n;\mathcal{O}) \to 0 \quad .$$

Since $\mathcal{A}'(\partial V) \to \mathcal{A}'(\bar{V})$ is injective it follows that $H_V^{n-1}(\mathbb{C}^n;\mathcal{O}) = 0$ and $H_V^n(\mathbb{C}^n;\mathcal{O}) \cong \mathcal{A}'(\bar{V})/\mathcal{A}'(\partial V)$.

Thus we have proved (i) for bounded sets V . This implies that $\mathcal{H}^p_{\mathbb{R}^n}(\mathcal{O}) = 0$ for $p \neq n$ and then (i) and (ii) follow from Theorem 1.3.6. (iii) follows from (ii) and the above, and (iv) is a consequence of (iii). To prove (v) notice that by (iii) $\mathcal{A}'(\bar{V}) \to \mathcal{B}(V)$ is onto and thus every hyperfunction on V extends to some neighborhood of $\bar{V}$. This implies the flabbiness of $\mathcal{B}$. □

Actually, (iii) could be used to give an alternative definition of hyperfunctions, without using cohomology, as follows (cf. Hörmander [b] or Schapira [a]): Let $V \subset \mathbb{R}^n$ be open and bounded then $\mathcal{A}'(\bar{V})/\mathcal{A}'(\partial V) = \mathcal{A}'(K)/\mathcal{A}'(K \setminus V)$ for all compact sets containing V . Call this space $B(V)$. If $V' \subset V$ is another open set it follows that there is a natural linear map from $B(V)$ to $B(V')$, which we call restriction. If $V \subset \mathbb{R}^n$ is open and unbounded, let $B(V) = 0$, then B is a presheaf on $\mathbb{R}^n$.

<u>Theorem 1.4.2</u> <u>The sheaves</u> $\mathcal{B}$ <u>and</u> $\Gamma^{\vee}B$ <u>are naturally isomorphic</u>.

<u>Proof</u>: See Schapira [a] Theorem 412 c . □

Notice that the definition of hyperfunctions immediately generalizes to an arbitrary n-dimensional oriented real analytic manifold M . Thus let X be a complex neighborhood of M , let $\mathcal{O}_X$ denote the sheaf of germs of holomorphic functions on X , and define:

$$\mathcal{B}_M = \mathcal{H}^n_M(\mathcal{O}_X) \quad .$$

One can prove the analog Theorem 1.4.1 in this general setting, too. In particular, if M is compact we have $\mathcal{B}(M) \cong \mathcal{A}'(M)$.

Let $V \subset \mathbb{R}^n$ be an open set. We will give the space $\mathcal{B}(V)$ an explicit description in terms of holomorphic functions, using Theorem 1.3.7. By the solution to Levi's problem (Grauert [a], Hörmander [a]) there exists a domain of holomorphy W in $\mathbb{C}^n$ such that $V = W \cap \mathbb{R}^n$. Put

$$W_j = \{z = (z_1,\ldots,z_n) \in W \mid \operatorname{Im} z_j \neq 0\}$$

for $j=1,\dots,n$, and let $\mathcal{W}' = \{W_1,\dots,W_n\}$ and $\mathcal{W} = \mathcal{W}' \cup \{W\}$. Then $(\mathcal{W},\mathcal{W}')$ is a relative covering of W modulo V, and $\mathcal{W}$ is acyclic for $\mathcal{O}$ since W is a domain of holomorphy. Therefore

$$(1.5) \qquad \mathcal{B}(V) \simeq \check{H}^n(\mathcal{W},\mathcal{W}';\mathcal{O})$$

and the latter space is easily computed from the definition. The result is:

$$(1.6) \qquad \mathcal{B}(V) \cong \frac{\mathcal{O}(\{z \in W \mid \forall j : \operatorname{Im} z_j \neq 0\}}{\sum_{k=1}^{n} \mathcal{O}(\{z \in W \mid \forall j \neq k : \operatorname{Im} z_j \neq 0\}} .$$

Since $\{z \in \mathbb{C}^n \mid \forall j : \operatorname{Im} z_j \neq 0\}$ has 2^n connected components, a hyperfunction f is represented by 2^n holomorphic functions. Intuitively f is the sum of 2^n boundary values of holomorphic functions (compare with the case $n=1$).

Using another relative covering one can in fact express the hyperfunctions on V by only $n+1$ holomorphic functions: Choose $n+1$ vectors $\eta_0,\dots,\eta_n$ in $\mathbb{R}^n \setminus 0$ such that $\mathbb{R}^n \setminus 0 = \bigcup_{j=0}^{n} E_j$, where $E_j = \{y \in \mathbb{R}^n \mid \langle y,\eta_j\rangle > 0\}$, and put $W_j = W \cap (V + \sqrt{-1}\, E_j)$. Let $\mathcal{W}' = \{W_0,\dots,W_n\}$ and $\mathcal{W} = \mathcal{W}' \cup \{W\}$, then $(\mathcal{W},\mathcal{W}')$ is an acyclic relative covering of W modulo V, and hence (1.5) holds again. It easily follows that

$$(1.7) \qquad \mathcal{B}(V) \cong \frac{\sum_{0\le j\le n} \mathcal{O}(W_0 \cap \dots \cap \widehat{W_j} \cap \dots \cap W_n)}{\sum_{0\le i<j\le n} \mathcal{O}(W_0 \cap \dots \cap \widehat{W_i} \cap \dots \cap \widehat{W_j} \cap \dots \cap W_n)} .$$

Let Γ_j denote the open convex cone in $\mathbb{R}^n$, given by

$$\Gamma_j = \{y \in \mathbb{R}^n \mid \forall k \neq j : \langle y,\eta_j\rangle > 0\} = E_0 \cap \dots \cap \widehat{E_j} \cap \dots \cap E_n$$

then a hyperfunction on V is represented by $n+1$ holomorphic functions on $W \cap (V + \sqrt{-1}\,\Gamma_j)$ for $j=0,\dots,n$, respectively.

Now, to give this concept of considering hyperfunctions as boundary values a rigorous definition, let $\Gamma \subset \mathbb{R}^n$ be an arbitrary open convex cone, and let $F \in \mathcal{O}(W \cap (V + \sqrt{-1}\,\Gamma))$. We will define the boundary value of F in $\mathcal{B}(V)$. Choose vectors $\eta_0,\dots,\eta_n$ as above, such that in addition $\Gamma_0 \subset \Gamma$, and let $\varepsilon = \pm 1$ be the orientation of $\eta_1,\dots,\eta_n$. Define $\varphi \in \Sigma_{0\le j\le n} \mathcal{O}(W \cap (V+\sqrt{-1}\,\Gamma_j))$ by

$\varphi = \varepsilon F$ on $W \cap (V + \sqrt{-1}\,\Gamma_0)$ and $\varphi = 0$ elsewhere, and let $f \in \mathcal{B}(V)$ be the equivalence class of φ by (1.7). Then f is independent of the choice of $\eta_0, \ldots, \eta_n$, and it is called the boundary value of F. It is denoted $b_\Gamma(F)$ (or $F(x + \sqrt{-1}\,\Gamma_0)$). Notice that $b_\Gamma(F)$ is in fact also independent of Γ, that is, if $\Gamma' \subset \Gamma$ is another open convex cone and $F' = F|_{W \cap (V + \sqrt{-1}\,\Gamma')}$, then $b_{\Gamma'}(F') = b_\Gamma(F)$.

In particular, we get a map $\mathcal{A}(V) \longrightarrow \mathcal{B}(V)$ by extending the analytic function to a holomorphic function on a neighborhood, and applying to this function b_Γ for an arbitrary open convex cone Γ.

Theorem 1.4.3 Let $V \subset \mathbb{R}^n$ be open and let W be a complex neighborhood which is a domain of holomorphy.

(i) For every hyperfunction $f \in \mathcal{B}(V)$ there exists a finite family $\Gamma_1, \ldots, \Gamma_I$ of open convex cones and a family of holomorphic functions $F_i \in \mathcal{O}(W \cap (V + \sqrt{-1}\,\Gamma_i))$ such that

$$f = \Sigma_{i=1}^{I} b_{\Gamma_i}(F_i) .$$

(ii) If $f = \Sigma_{i=1}^{I} b_{\Gamma_i}(F'_i)$ with $F'_i \in \mathcal{O}(W \cap (V + \sqrt{-1}\,\Gamma_i))$ is another such representation (with the same family of cones Γ_i) then there exist functions $G_{ij} \in \mathcal{O}(W \cap (V + \sqrt{-1}\,\Gamma_{ij}))$, where Γ_{ij} denotes the convex hull of $\Gamma_i \cup \Gamma_j$, such that

$$F_i - F'_i = \Sigma_{j=1}^{I} G_{ij}|_{W \cap (V + \sqrt{-1}\,\Gamma_i)}$$

for $i = 1, \ldots, I$, and $G_{ij} = -G_{ji}$.

Proof: (i) is immediate from (1.7). (ii) is a version of Martineau's edge of the wedge theorem, see Martineau [b]. □

Notice that (ii) implies that $F \longrightarrow b_\Gamma(F)$ is an injection of $\mathcal{O}(W \cap (V + \sqrt{-1}\,\Gamma))$ into $\mathcal{B}(V)$. In particular, it follows that the map $\mathcal{A}(V) \longrightarrow \mathcal{B}(V)$ defined above is injective.

More generally, the space $\mathcal{D}'(V)$ of distributions on V can be injected into $\mathcal{B}(V)$ in a natural way. As in the case of one variable, this follows from Theorem 1.4.1, but one can also give the injection an interpretation in terms of boundary values of holomorphic functions satisfying certain growth conditions, cf. Hörmander [b] Theorems 8.4.15 and 9.3.3(iii).

Let $P(x,D) = \sum_{|\alpha| \leq m} a_\alpha(x) D_x^\alpha$ be a differential operator with real analytic coefficients on V. Then P extends analytically to $P(z,D) = \sum_{|\alpha| \leq m} a_\alpha(z) D_z^\alpha$ on some neighborhood W of V, and P thus defines a sheaf homomorphism $\mathcal{O} \longrightarrow \mathcal{O}$ on W. Then P acts on $\mathcal{B}(V) = H_V^n(W;\mathcal{O})$ by the induced homomorphism. In terms of the notation from Theorem 1.4.4(i) this means that if $f = \Sigma_{i=1}^I b_{\Gamma_i}(F_i)$ then $Pf = \Sigma_{i=1}^I b_{\Gamma_i}(PF_i)$, and by (ii) of that theorem, Pf is independent of the chosen representation of f. In particular, we thus know how to <u>differentiate</u> hyperfunctions.

By Theorem 1.4.1(iv) it is straightforward to define the <u>integral</u> of a hyperfunction with compact support, in fact, we just have to evaluate the corresponding analytic functional on the constant function 1.

More generally, if $M = M_1 \times M_2$ where M_1 and M_2 are real analytic manifolds and M_2 is compact, then it is possible (cf. Sato et al. [a] p. 295, Theorem 2.3.1 or Bony [a] p. 33 ff) to define the integral over M_2 of a hyperfunction f on M. The integral is then a hyperfunction on M_1:

$$\int_{M_2} f(x_1,x_2)dx_2 \in \mathcal{B}(M_1) .$$

Let $V' \subset \mathbb{R}^{n'}$ and $V \subset \mathbb{R}^n$ be open sets, and let $\varphi : V' \longrightarrow V$ be a real analytic map, which we assume is a submersion, that is, the differential $d\varphi_x : \mathbb{R}^{n'} \longrightarrow \mathbb{R}^n$ is surjective at each $x \in V'$ (so in particular $n \leq n'$). Let $f \in \mathcal{B}(V)$ be a hyperfunction on V. We will define the <u>composite</u> $f \circ \varphi \in \mathcal{B}(V')$. We extend φ to an analytic map Φ from a neighborhood of V' in $\mathbb{C}^{n'}$ to a neighborhood of V in $\mathbb{C}^n$. According to Theorem 1.4.4(i) we write f as a sum of boundary values $f = \Sigma_i b_{\Gamma_i}(F_i)$, where Γ_i are open convex cones of $\mathbb{R}^n$ and $F_i \in \mathcal{O}(W \cap (V + \sqrt{-1}\,\Gamma_i))$. Let $x_0 \in V'$ and let Γ_i' be an open convex cone of $\mathbb{R}^{n'}$ whose closure is contained in $d\varphi_{x_0}^{-1}(\Gamma_i)$. Then there is a neighborhood W' of x_0 in $\mathbb{C}^{n'}$, such that

$$\Phi(W' \cap (\mathbb{R}^{n'} + \sqrt{-1}\,\Gamma_i')) \subset W \cap (V + \sqrt{-1}\,\Gamma_i) .$$

We can then define, in a neighborhood of x_0:

$$f \circ \varphi = \Sigma_i b_{\Gamma_i}(F_i \circ \Phi) .$$

It follows from Theorem 1.4.3(ii) that this definition of $f \circ \varphi$ is independent of all choices made.

Example 1.4.4 Hyperfunctions on $\mathbb{R}^2$

1) Let $f_1, f_2 \in \mathcal{B}(\mathbb{R})$ be hyperfunctions on $\mathbb{R}$ with defining functions $F_1, F_2 \in \mathcal{O}(\mathbb{C}\setminus\mathbb{R})$. Define $F \in \mathcal{O}(\mathbb{C}\setminus\mathbb{R} \times \mathbb{C}\setminus\mathbb{R})$ by

$$F(z_1, z_2) = F_1(z_1)F_2(z_2)$$

then by (1.6) F determines a hyperfunction on $\mathbb{R}^2$, which we denote $f(x) = f_1(x_1)f_2(x_2)$. Thus, we can define $\delta(x_1, x_2) = \delta(x_1)\delta(x_2)$, where $\delta(x_i)$ is Dirac's deltafunction on $\mathbb{R}$ as defined in Section 1.1. Also, if $\lambda_i \in \mathbb{C}\setminus\{-1,-2,\dots\}$ for $i = 1,2$, we can define $x_+^\lambda \in \mathcal{B}(\mathbb{R}^2)$ by $x_+^\lambda = (x_1)_+^{\lambda_1}(x_2)_+^{\lambda_2}$.

2) Let $f \in \mathcal{B}(\mathbb{R})$, then $f(x_1 - x_2) \in \mathcal{B}(\mathbb{R}^2)$ makes sense by the definition of the composite. Thus, for instance, $\delta(x_1 - x_2)$ is a well defined hyperfunction on $\mathbb{R}^2$, whose support is the diagonal $\{x_1 = x_2\}$.

Example 1.4.5 Convolution

Let $f_1, f_2 \in \mathcal{B}(\mathbb{R})$ and assume that f_2 has compact support. By the preceding example $f_1(x_1 - x_2)f_2(x_2)$ makes sense as a hyperfunction on $\mathbb{R}^2$ with support in $\mathbb{R} \times \operatorname{supp} f_2$. Integrating it over x_2 we get a hyperfunction on $\mathbb{R}$, which we denote $f_1 * f_2$:

$$f_1 * f_2(x_1) = \int_{\mathbb{R}} f_1(x_1 - x_2)f_2(x_2)dx_2 .$$

This definition is easily generalized to an arbitrary Lie group instead of $\mathbb{R}$.

Example 1.4.6 Let G be a Lie group. For $f \in \mathcal{B}(G)$ and $g \in G$ the translated hyperfunctions $f(g\cdot\ )$ and $f(\cdot g)$ on G are defined by composition of f with translation. If H is a Lie subgroup we can then identify $\mathcal{B}(G/H)$ naturally with the space

$$\{f \in \mathcal{B}(G) \mid f(\cdot h) = f \quad \forall h \in H\}$$

by composition of $f \in \mathcal{B}(G/H)$ with the projection $G \longrightarrow G/H$. If moreover H is compact, then we can define a surjection

$$\mathcal{B}(G) \ni f \longrightarrow f^H \in \mathcal{B}(G/H)$$

by integrating the function $(g,h) \longrightarrow f(gh)$ over H.

1.5. The singular spectrum and microfunctions

In this section we will introduce the sheaf $\mathcal{C}$ of microfunctions. Microfunctions are used for the detailed study of the local structure of solutions to partial differential equations on analytic manifolds by means of the cotangent bundle, so-called micro-local analysis.

Let M be a real analytic oriented manifold and let S^*M be the cotangent sphere bundle, that is

$$S^*M = (T^*M \setminus \{0\})/]0,\infty[\ .$$

It is convenient to multiply all elements of S^*M with $\sqrt{-1}$ and instead consider $\sqrt{-1}\,S^*M$. We will use the notation $(x, \sqrt{-1}\,\xi\infty)$ for points of $\sqrt{-1}\,S^*M$, where $x \in M$ and $\xi \in T^*_xM \setminus \{0\}$.

For convenience, and since our business is local, we will assume that M is an open subset of $\mathbb{R}^n$ and hence make the identification $T^*M \simeq M \times \mathbb{R}^n$.

Let $(x_0, \sqrt{-1}\,\xi_0\infty) \in \sqrt{-1}\,S^*M$. A hyperfunction f defined on a neighborhood of x_0 is called micro-analytic at $(x_0, \sqrt{-1}\,\xi_0\infty)$ if f can be expressed as a sum of boundary values $f = \Sigma_j b_{\Gamma_j}(F_j)$ in a neighborhood of x_0 with $\langle \eta, \xi_0 \rangle < 0$ for all $\eta \in \cup_j \Gamma_j$. The set of points in $\sqrt{-1}\,S^*M$ where f is not micro-analytic is called the singular spectrum (or analytic wave front set) of f and is denoted $SS(f)$ (or $WF_A(f)$).

Notice that f is analytic in a neighborhood of x_0 if and only if $SS(f) \cap \sqrt{-1}\,S^*_{x_0}M = \emptyset$.

Let $\Gamma \subset \mathbb{R}^n$ be an open convex cone and $V \subset \mathbb{R}^n$ an open set. Consider the following property of an open set $U \subset \mathbb{C}^n$: For every open set $V' \subset \mathbb{R}^n$ with $\overline{V'} \subset V$ and every open convex cone Γ' with $\overline{\Gamma'} \subset \Gamma$ there exists $\varepsilon > 0$ such that

$$(V' + \sqrt{-1}\,\Gamma') \cap \{z \in \mathbb{C}^n \mid |\mathrm{Im}\, z| < \varepsilon\} \subset U \ .$$

We call Γ-holomorphic near V any holomorphic function F defined on an open set $U \subset \mathbb{C}^n$ with this property. By $\mathcal{O}(V + \sqrt{-1}\,0\Gamma)$ we denote the space of functions which are Γ-holomorphic near V (identifying two Γ-holomorphic functions if they are identical on the intersection of their domains of definition with a neighborhood of V). For $F \in \mathcal{O}(V + \sqrt{-1}\,0\Gamma)$ we define the boundary value $b_\Gamma(F)$ of F by

$$b_\Gamma(F)|_{V'} = b_{\Gamma'}(F|_{(V' + \sqrt{-1}\,\Gamma') \cap \{z \in \mathbb{C}^n \mid |\mathrm{Im}\, z| < \varepsilon\}})$$

for any V' and Γ' as above (the boundary value on the right hand side makes sense by the preceding section).

The dual cone $\Gamma^0 \subset \mathbb{R}^n$ of Γ is defined by $\Gamma^0 = \{\xi \in \mathbb{R}^n \mid \langle \xi, \eta \rangle \geq 0 \quad \forall \eta \in \Gamma\}$.

Proposition 1.5.1 *Let* $\Gamma \subset \mathbb{R}^n$ *be a convex open cone and* $V \subset \mathbb{R}^n$ *an open set. Denote by* $V \times \sqrt{-1}\,\Gamma^0\infty$ *the set of points* $(x, \sqrt{-1}\,\xi\infty) \in S^*M$ *with* $\xi \in \Gamma^0$. *Let* $f \in \mathcal{B}(V)$. *Then*

$$SS(f) \subset V \times \sqrt{-1}\,\Gamma^0\infty$$

if and only if f *is the boundary value of some* Γ*-holomorphic function.*

Proof: See Sato et al. [a] p. 285, Proposition 1.5.4 or Hörmander [b], Theorem 9.3.4. □

Let $U \subset \sqrt{-1}\, S^*M$ be an open set. The space $\mathcal{C}(U)$ of *microfunctions* on U is defined by

$$\mathcal{C}(U) = \mathcal{B}(M)/\{f \in \mathcal{B}(M) \mid SSf \cap U = \emptyset\} \ . \tag{1.8}$$

With the obvious map of restrictions $U \longrightarrow \mathcal{C}(U)$ is a presheaf on $\sqrt{-1}\, S^*M$. Let $\pi : \sqrt{-1}\, S^*M \longrightarrow M$ be the base map.

Theorem 1.5.2 *The presheaf* $U \longrightarrow \mathcal{C}(U)$ *has the localization property. The corresponding sheaf, also denoted* $\mathcal{C}$, *is flabby.*

Proof: In Sato et al. [a] p. 276, Definition 1.3.3 (cf. also Sato[d]) a more intrinsic definition of a sheaf $\mathcal{C}$ on $\sqrt{-1}\, S^*M$ is given: $\mathcal{C} = \mathcal{H}^n_{\sqrt{-1}\, S^*M}(\pi^{-1}\mathcal{O})^a$, where a is the antipodal map $\sqrt{-1}\, S^*M \longrightarrow \sqrt{-1}\, S^*M$. In loc. cit. p. 473, Corollary 2.1.5 it is proved, using Theorem 1.5.6 below (the proof of which involves the machinery of Section 1.6), that this sheaf is flabby. From loc. cit. p. 284, Theorem 1.5.3 it follows that the sections over U of this sheaf are given by (1.8). □

For each open set $V \subset M$, let $sp : \mathcal{B}(V) \longrightarrow \mathcal{C}(\pi^{-1}V)$ be the quotient map, then we have for $f \in \mathcal{B}(M)$:

$$SSf = \text{supp}(sp\ f) \ .$$

Proposition 1.5.3 Let $f \in \mathcal{B}(V)$ and let $\Gamma_1,\dots,\Gamma_I \subset \mathbb{R}^n$ be open convex cones such that $SS(f) \subset V \times \sqrt{-1}(\cup_{i=1}^I \Gamma_i^0)^\infty$. Then f can be written in the form

$$f = \Sigma_{i=1}^I b_{\Gamma_i}(F_i)$$

for some $F_i \in \mathcal{O}(V + \sqrt{-1}\, 0\, \Gamma_i)$ $(i = 1,\dots,I)$.

Proof: Since $\operatorname{supp} sp(f) \subset \bigcup_{i=1}^{I}(V \times \sqrt{-1}\,\Gamma_i^0\infty)$ and $\mathcal{C}$ is flabby we can write $sp(f) = \Sigma_{i=1}^I v_i$ for some microfunctions $v_1,\dots,v_I$ with $\operatorname{supp} v_i \subset V \times \sqrt{-1}\,\Gamma_i^0\infty$. Let $f_i \in \mathcal{B}(V)$ be an arbitrary hyperfunction with $sp\, f_i = v_i$. Then $sp(f - \Sigma_{i=1}^I f_i) = 0$, so $f - \Sigma_{i=1}^I f_i \in \mathcal{A}(V)$. Altering f_1 if necessary we may assume $f = \Sigma_{i=1}^I f_i$. Then the proposition follows from Proposition 1.5.1. □

Similarly, the flabbiness of $\mathcal{C}$ implies an "edge of the wedge" theorem, analogous to Theorem 1.4.3(ii) (cf. Morimoto [a], Theorem 16).

Let N be a real analytic submanifold of M. Then it is possible to define the restriction to N of a hyperfunction on M, provided its singular spectrum satisfies a certain condition.

For simplicity, let $M = \mathbb{R}^n$ and suppose N is given by $x_{n+1} = \dots = x_m = 0$. Let $T_N^*M = \{(x, \xi) \in T^*M \mid x \in N, \xi_1 = \dots = \xi_n = 0\}$ and let $\sqrt{-1}\, S_N^*M$ denote the corresponding subset of $\sqrt{-1}\, S^*M$. Let $f \in \mathcal{B}(M)$ and assume that $SSf \cap \sqrt{-1}\, S_N^*M = \emptyset$.

Let $x_0 \in N$. The condition on $SS(f)$ implies that there are open convex cones $\Gamma_1,\dots,\Gamma_I$ in $\mathbb{R}^m$ such that $\Gamma_i^0 \cap \{\xi \mid \xi_1 = \dots = \xi_n = 0\} = \emptyset$ for all i, and

$$SS(f) \subset \bigcup_{i=1}^{I}(V \times \sqrt{-1}\,\Gamma_i^0\infty) .$$

in a neighborhood V of x_0 in M. By the preceding proposition there are functions $F_i \in \mathcal{O}(V + \sqrt{-1}\, 0\, \Gamma_i)$ such that

$$f = \Sigma_{i=1}^I b_{\Gamma_i}(F_i)$$

Let $\Gamma_i' = \Gamma_i \cap \{\xi \mid \xi_{n+1} = \dots = \xi_m = 0\}$, then Γ_i' is a nonempty open convex cone in $\mathbb{R}^n$ $(i = 1,\dots,I)$. Let $F_i \in \mathcal{O}(V \cap N + \sqrt{-1}\, 0\, \Gamma_i')$ be the restriction of F_i.

<u>Proposition 1.5.4</u> <u>The hyperfunction</u> $f|_N$ <u>on</u> N <u>given by</u>

$$f|_N = \Sigma_{i=1}^{I} b_{\Gamma_i'}(F_i)$$

<u>in a neighborhood of</u> x_0 <u>in</u> N <u>is independent of the choices of</u> $F_1,\ldots,F_I$ <u>and</u> $\Gamma_1,\ldots,\Gamma_I$. <u>Moreover</u>

$$SS(f|_N) \subset \{(x,\sqrt{-1}\,\xi\infty) \in \sqrt{-1}\,S^*M \mid x \in N,\ \exists\,\eta \in \mathbb{R}^{m-n}:(x,0,\sqrt{-1}(\xi,\eta)\infty) \in SS(f)\}.$$

<u>Proof</u>: The proof is straightforward by the edge of the wedge theorem. □

By a similar argument, one can define the product of two hyperfunctions f and g , provided $SSf \cap (SSg)^a = \emptyset$ where $(SSg)^a = \{(x,-\sqrt{-1}\,\xi\infty) \mid (x,\sqrt{-1}\,\xi\infty) \in SSg\}$ (Sato et al. [a] p. 297, Corollary 2.4.2).

In section 1.4 we mentioned that if $M = M_1 \times M_2$ with M_2 compact, then the integral $h = \int_{M_2} f(x_1,x_2)dx_2 \in \mathcal{B}(M_1)$ is defined for $f \in \mathcal{B}(M)$. In the references given for this it is also proved that

$$SS\,h \subset \{(x_1,\sqrt{-1}\,\xi\infty) \in \sqrt{-1}\,S^*M_1 \mid \exists\, x_2 \in M_2 : (x_1,x_2,\sqrt{-1}(\xi,0)\infty) \in SSf\}.$$

It is easily seen that this has the consequence that one can <u>integrate microfunctions</u> too. In fact, if u is a microfunction defined in a neighborhood of $(x_1,x_2,\sqrt{-1}(\xi,0)\infty) \in \sqrt{-1}\,S^*M$ for all $x_2 \in M_2$ and for a given point $(x_1,\sqrt{-1}\,\xi\infty)$ in $\sqrt{-1}\,S^*M_1$, then if $u = spf$ in a neighborhood of these points we can define $\int u\,(x_1,x_2)dx_2 = sp\,h$ where h is the integral of f . It follows that this is independent of the choice of f in a neighborhood of $(x_1,\sqrt{-1}\,\xi\infty)$. (Sato et al. [a] p. 295, Theorem 2.3.1).

Let $P(x,D) = \Sigma_{|\alpha| \le m} a_\alpha(x)D_x^\alpha$ be a differential operator with real analytic coefficients on M . From Proposition 1.5.3 it follows that

$$SS(Pf) \subset SS(f) \tag{1.9}$$

for $f \in \mathcal{B}(M)$, and hence by (1.8) <u>P operates on $\mathcal{C}$ (U)</u> for every open set $U \subset \sqrt{-1}\,S^*M$.

The inclusion (1.9) is in general not an equality, but if P is elliptic, it is. Indeed, let

$$P_m(x,\xi) = \Sigma_{|\alpha|=m} a_\alpha(x)\xi^\alpha$$

for $x \in M$, $\xi \in \mathbb{R}^n$, the principal symbol, and

$$\text{char } P = \{(x, \sqrt{-1}\,\xi\infty) \in S^*M \mid P_m(x,\xi) = 0\}$$

the characteristic variety of P.

Theorem 1.5.5 Let P be a differential operator with real analytic coefficients on M and let $f \in \mathcal{B}(M)$. Then

$$SS(f) \subset SS(Pf) \cup \text{char } P \;.$$

Proof: The proof given by Sato involves the theory of the sheaf $\mathcal{C}$, in fact the theorem is an immediate corollary of Theorem 1.5.6 below. Another proof has been given by Bony and Schapira [a] (cf. also Hörmander [b], Theorem 9.5.1.) □

By definition, P is elliptic if char P = ∅, and thus in this case SS(Pf) = SS(f). In particular, every hyperfunction solution f to an elliptic analytic differential equation Pf = 0 is an analytic function.

Theorem 1.5.6 Let P be a differential operator with real analytic coefficients on M. Then the endomorphism of $\mathcal{C}$ induced by P is bijective on $\sqrt{-1}\, S^*M \setminus \text{char } P$.

Proof: This is proved in Sato et al. [a] by actual construction of the inverse operator, which is a micro-differential operator. See the next section. □

Finally in this section we mention the following result, called "Holmgren's uniqueness principle", which is proved in Sato et al. [a] (p. 471, Proposition 2.1.3) as a corollary to Theorem 1.5.5.

Proposition 1.5.7 Let f be a hyperfunction defined in a neighborhood of 0 in $\mathbb{R}^n$ and supported on the set $\{x_n \geq 0\}$. If the points

$(0, \sqrt{-1}(0,\ldots,0,\pm 1)\infty)$ are not in SS(f), then f vanishes in a neighborhood of 0.

1.6. Micro-differential operators

As mentioned, the proof of Theorem 1.5.6 involved construction of an inverse operator to P. This operator is what Sato, Kashiwara and Kawai in [a] call a pseudo-differential operator in analogy with the operators of Kohn and Nirenberg. In recent literature the term micro-differential operator is used.

Let X be a complex analytic manifold and P^*X its cotangent projective bundle, i.e., $P_z^*X = (T_z^*X \setminus \{0\})/(\mathbb{C}\setminus\{0\})$ for $z \in X$, and let $\pi : T^*X \setminus \{0\} \longrightarrow P^*X$ be the canonical projection. As before, we are only interested in local matters, and we assume that X is an open subset of $\mathbb{C}^n$ and identify T^*X with $X \times \mathbb{C}^n$.

Let $U \subset P^*X$ be an open set and m an integer. The space $\mathcal{E}^{(m)}(U)$ of micro-differential operators of order at most m on U is by definition the space of formal expressions

$$(1.10) \qquad P(z,\zeta) = \sum_{k=-\infty}^{m} P_k(z,\zeta)$$

such that the following (i) - (ii) hold:

(i) For each $k \leq m$, P_k is a holomorphic function in $(z,\zeta) \in \pi^{-1}(U)$, which for fixed $z \in X$ is homogeneous of degree k in $\zeta \in T_z^*X$, that is,

$$P_k(z, \lambda\zeta) = \lambda^k P_k(z,\zeta)$$

for all $\lambda \in \mathbb{C}\setminus\{0\}$.

(ii) For every compact subset $K \subset \pi^{-1}(U)$ there exists a constant $C_K > 0$ such that

$$|P_k(z,\zeta)| \leq (C_K)^{|k|}\,|k|!$$

for all $(z,\zeta) \in K$ and $k \leq m$.

Obviously, the presheaf $U \longrightarrow \mathcal{E}^{(m)}(U)$ has the localization property, and thus $\mathcal{E}^{(m)}(U)$ consists of the sections over U of a sheaf $\mathcal{E}^{(m)}$ on P^*X. We regard $\mathcal{E}^{(m-1)}(U)$ as a subspace of $\mathcal{E}^{(m)}(U)$ for all $m \in \mathbb{Z}$, and define $\mathcal{E}(U) = \bigcup_{m\in\mathbb{Z}} \mathcal{E}^{(m)}(U)$.

In this way we obtain a sheaf $\mathcal{E}$, the <u>sheaf of micro-differential operators</u> (of finite order).

When $P \in \mathcal{E}^{(m)}(U)$ is given by (1.10) we denote $\sigma_m(P) = P_m \in \mathcal{O}(\pi^{-1}(U))$. Then the space $\mathcal{E}^{(m)}(U)/\mathcal{E}^{(m-1)}(U)$ is isomorphic via σ_m to the space of holomorphic functions on $\pi^{-1}(U)$ homogeneous of degree m with respect to ζ . If $\sigma_m(P)$ is not identically zero it is called the principal symbol of P , and m is the order of P .

Notice that the differential operators with holomorphic coefficients on X can be regarded as elements of $\mathcal{E}(P^*X)$, such that the operator $\Sigma_{|\alpha|\leq m} a_\alpha(z)D_z^\alpha$ corresponds to the element (1.10) of $\mathcal{E}(P^*X)$ where $P_k(z,\zeta) = \Sigma_{|\alpha|=k} a_\alpha(z)\zeta^\alpha$ for $k \geq 0$ and $P_k(z,\zeta) = 0$ otherwise (in fact, if $n \geq 2$ these are the only elements of $\mathcal{E}(P^*X)$).

Condition (ii) above can be expressed in a more technical, but often convenient manner as follows. Let $P(z,\zeta) = \Sigma_{k=-\infty}^{m} P_k(z,\zeta)$ satisfy (i) above. For each $t > 0$ and each open and bounded subset ω of $\pi^{-1}(U)$ with $\bar{\omega} \subset \pi^{-1}(U)$ we define

$$(1.11) \quad N_m^\omega(P;t) = \Sigma_{k=0}^{\infty} \sum_{\alpha,\beta} \frac{2(2n)^{-k}k!}{(k+|\alpha|)!(k+|\beta|)!} \sup_\omega |D_z^\alpha D_\zeta^\beta P_{m-k}| \, t^{2k+|\alpha|+|\beta|}$$

(cf. Boutet de Monvel [a]). Then we have for $(z_0,\zeta_0) \in \pi^{-1}(U)$

<u>Lemma 1.6.1</u> <u>Condition</u> (ii) <u>holds in a neighborhood of</u> (z_0,ζ_0) <u>if and only if</u> $N_m^\omega(P,t) < \infty$ <u>for some</u> $t > 0$ <u>and some neighborhood</u> ω <u>of</u> (z_0,ζ_0).

<u>Proof:</u> See Björk [a] Ch. 4, Lemma 3.2 and 3.3. □

The sheaf $\mathcal{E}$ can be studied algebraically. Extending the Leibniz rule for composition of differential operators we define for two micro-differential operators P and Q on U

$$(1.12) \qquad (PQ)_k(z,\zeta) = \Sigma \frac{1}{\alpha!} D_\zeta^\alpha P_j(z,\zeta) \, D_z^\alpha Q_\ell(z,\zeta)$$

where the sum ranges over all multiindices α and all integers j and ℓ such that $j + \ell - |\alpha| = k$. Since P and Q have finite order, the sum is finite.

Proposition 1.6.2 The expression (1.12) defines a micro-differential operator PQ on U. With this composition rule as product, $\mathcal{E}$ is a sheaf of rings. If $P \in \mathcal{E}^{(m)}$ and $Q \in \mathcal{E}^{(m')}$ then $PQ \in \mathcal{E}^{(m+m')}$ and $\sigma_{m+m'}(PQ) = \sigma_m(P)\sigma_{m'}(Q)$. Also for all $t > 0$ and $\overline{\omega} \subset \pi^{-1}(U)$, ω open and bounded, we have

$$(1.13) \qquad N^{\omega}_{m+m'}(PQ;t) \leq N^{\omega}_m(P;t)\, N^{\omega}_m(Q;t) \quad .$$

Proof: It is obvious that $PQ = \sum_{k=-\infty}^{m+m'} (PQ)_K$ satisfies condition (i) of the definition of micro-differential operators. By Lemma 1.6.1 we only have to prove (1.13), the remaining claims being immediate. See Björk [a] Ch. 4, Theorem 3.4. □

Let P be a micro-differential operator. With the product defined it makes sense to apply to P a polynomial in one variable with complex coefficients: If $f(s) = \Sigma_{j=0}^m a_j s^j$ $(a_0,\ldots,a_m \in \mathbb{C})$ we define $f(P) = \Sigma_{j=0}^m a_j P^j$. This kind of operation on P can in fact be extended to all functions f holomorphic in a neighborhood of 0 in $\mathbb{C}$, provided P is sufficiently nice. Thus let $f(s) = \Sigma_{j=0}^{\infty} a_j s^j$ be convergent in a neighborhood of 0, and let P be of order ≤ 0 in a neighborhood of $(z_0, \zeta_0\infty) \in P^*X$ and assume that $P_0(z_0,\zeta_0) = 0$.

Proposition 1.6.3 The expression $S = \Sigma_{j=0}^{\infty} a_j P^j$ defines a micro-differential operator of order ≤ 0 near $(z_0,\zeta_0\infty)$, in the sense that for each $k \leq 0$ the series $S_k = \Sigma_{j=0}^{\infty} a_j (P^j)_k$ converges and the sum $\Sigma_{k=-\infty}^{0} S_k$ is a micro-differential operator.

Proof: Choose $\varepsilon > 0$ such that $\Sigma_{j=0}^{\infty} |a_j|\varepsilon^j < \infty$. Since $N_0^{\omega}(P;t) \longrightarrow 2\sup_{\omega}|P_0|$ as $t \to 0$, we can choose a neighborhood ω of (z_0,ζ_0) and $t > 0$ such that $N_0^{\omega}(P;t) < \varepsilon$. Then by (1.13) $N_0^{\omega}(P^j;t) < \varepsilon^j$ and in particular

$$\frac{2(2n)^{-k}}{k!} \sup_{\omega} |(P^j)_k| t^{2k} < \varepsilon^j$$

which implies that $\Sigma_{j=0}^{\infty} a_j (P^j)_k$ converges uniformly on ω. Let $S_k = \Sigma_{j=0}^{\infty} a_j (P^j)_k$, then S satisfies Condition (i) and furthermore we get

$$N_0^{(\omega)}(S;t) \leq \Sigma_{j=0}^{\infty} |a_j| N_0^{(\omega)}(P^j;t) < \infty$$

and thus $S \in \mathcal{E}^{(0)}$ in a neighborhood of $(z_0, \zeta_0 \infty)$ by Lemma 1.6.1. □

Let $P \in \mathcal{E}^{(m)}(U)$ be a micro-differential operator and let $(z_0, \zeta_0 \infty) \in U$ be a point where P is defined. We say that P is <u>elliptic</u> at $(z_0, \zeta_0 \infty)$ if the principal symbol $\sigma_m(P)$ of P does not vanish at (z_0, ζ_0). It is a very important property of the sheaf of rings $\mathcal{E}$ that one can divide by elliptic elements:

<u>Theorem 1.6.4</u> <u>Let</u> P <u>be a micro-differential operator which is defined and elliptic in the open set</u> $U \subset P^*X$. <u>There exists a unique micro-differential operator</u> $Q \in \mathcal{E}(U)$ <u>such that</u> $PQ = QP = 1$ <u>on</u> U.

<u>Proof</u>: Let $(z_0, \zeta_0 \infty) \in U$. First we construct a left inverse to P in a neighborhood of $(z_0, \zeta_0 \infty)$. Let m be the order of P. Exchanging P with $D_{z_1}^{-m} P$ we may assume that $m = 0$. Put $R = 1 - \sigma_0(P)^{-1} P$ then $R \in \mathcal{E}^{(-1)}(U)$. For $s \in \mathbb{C}$, $|s| < 1$ let $f(s) = (1-s)^{-1} = \Sigma_{j=0}^{\infty} s^j$ then by Proposition 1.6.3 $f(R)$ is a micro-differential operator near $(z_0, \zeta_0 \infty)$ and $f(R)(1-R) = 1$. Taking $Q = \sigma_0(P) f(R)$ we get that $QP = 1$. Similarly one constructs a right inverse to P in a neighborhood of $(z_0, \zeta_0 \infty)$. The theorem easily follows. □

The crucial fact about micro-differential operators on X is that if X is the complexification of a real analytic oriented manifold M, then they operate on the sheaf $\mathcal{C}$ of microfunctions on $\sqrt{-1}\, S^*M$. However, even the very definition of how this operation takes place is quite complicated, and we shall only be able to sketch it very briefly. For details we refer to Sato et al. [a] Ch. II 1 (see also Kashiwara [a]).

In case of $M = \mathbb{R}$ the definition is reasonably simple: First an auxiliary hyperfunction φ_λ for $\lambda \in \mathbb{C}$ is defined as follows: For $z \in \mathbb{C} \setminus \mathbb{R}$ let

$$\Phi_\lambda(z) = \begin{cases} \Gamma(\lambda)\exp(-\lambda \operatorname{Log}(-z)) & \text{if } \lambda \notin \{-1,-2,\dots\} \\ -\dfrac{z^{-\lambda}}{(-\lambda)!} \operatorname{Log}(-z) & \text{if } \lambda \in \{-1,-2,\dots\} \end{cases}$$

and let $\varphi_\lambda = [\Phi_\lambda]$. Then $\frac{d}{dz} \varphi_\lambda = \varphi_{\lambda+1}$ for all $\lambda \in \mathbb{C}$. By

Cauchy's integration formula, one checks that for $f \in \mathcal{B}(\mathbb{R})$ with compact support:

$$\left(\frac{d}{dx}\right)^j f(x) = \frac{1}{2\pi}\int_{\mathbb{R}} (\sqrt{-1})^j \ \varphi_{1+j}(x-y)f(y)dy$$

when $j \geq 0$. This suggests using φ_{1+j} as a kernel to define $\left(\frac{d}{dx}\right)^j$ for $j < 0$ too. Let $f \in \mathcal{C}(U \times \{\sqrt{-1}\,\infty\})$ where $U \subset \mathbb{R}$ is open and bounded and let $\tilde{f} \in \mathcal{B}(U)$ be in the class of f . Let $P(z,\zeta) = \Sigma_{j=-\infty}^{m} a_j(z)\zeta^j$ be a micro-differential operator defined on a complex neighborhood of U and define

$$Pf = sp(\frac{1}{2\pi}\int_{\mathbb{R}} \Sigma_{j=-\infty}^{m} (\sqrt{-1})^j \ a_j(x)\varphi_{1+j}(x-y)\tilde{f}(y)dy)$$

then one can prove from Condition (ii) that the sum converges and moreover that this expression for Pf is in fact well defined. Finally, one can prove that $(PQ)f = P(Qf)$.

In the higher dimensional case a similar (but of course more complicated) method can be applied. In that way $\mathcal{C}$ becomes a <u>sheaf of $\mathcal{E}$ modules</u>. Combining this with Theorem 1.6.4, Theorem 1.5.6 immediately follows.

In Bony [a] a somewhat different construction of the action of $\mathcal{E}$ on $\mathcal{C}$ is given, based upon solutions to certain Cauchy problems (cf. also Björk [a] p. 315-317).

1.7 Notes

Hyperfunctions were introduced by M. Sato in 1958 ([a] I and II). In dimension one the idea of representing generalized functions as the "jump" between analytic functions above and below the real axis was earlier considered by T. Carleman. Also, in dimension one, analytic functionals had been studied by L. Fantappié and G. Köthe (see Lützen [a] p. 189-192, and the references given there). The introduction given here owes much to Cerezo et al. [a] and Komatsu [c] (and to Jørgensen [a] in danish).

As basic references to sheaves and their cohomology we mention Hirzebruch [a], Grauert and Remmert [a], Bredon [a]. We have presupposed some basic knowledge of homology theory on part of the reader as for instance is offered by Hilton and Stammbach [a]. The theory of local cohomology goes back to Grothendieck [a] (see also

Hartshorne [a]) and Sato[a]II. This theory also enters in representation theory (cf. Zuckerman [a]). For the purpose of hyperfunctions, thorough treatments are given in Komatsu [d] and Schapira [a].

Theorem 1.4.2 is due to M. Sato and Theorem 1.4.3 to A. Martineau ([a]). The approach to hyperfunctions via analytic functionals is used in Schapira [a] and in Hörmander [b] Ch. 9. For the edge of the wedge theorem and the representation of hyperfunctions as boundary values, see Morimoto [a].

Microfunctions were invented by M. Sato in 1969 [b], and the theory of these and micro-differential operators was further developed by Sato in collaboration with T. Kawai and M. Kashiwara. The outcome of this collaboration was the celebrated "SKK", Sato et al. [a]. Theorem 1.5.6 is called the "fundamental theorem of Sato" ([c]).

Holmgren's uniqueness (Proposition 1.5.7) was first proved for hyperfunctions by Schapira ([c]).

The algebraic theory of the sheaf $\mathcal{E}$ is given a thorough exposition in Björk [a], based on "SKK". Theorem 1.6.4 is Theorem 2.1.1 of "SKK" p. 356, and Proposition 1.6.3 is given on p. 358 of "SKK".

The theory of micro-differential operators goes much deeper than we have been able to indicate here. Many of the results from "SKK" are explained in Bony [a]. For further results, see Kashiwara [b].

Finally, here is a list of introductory texts on these topics: Cerezo et al. [a], Kashiwara [a], Kashiwara and Kawai [a], Kawai [a], and Miwa et al. [a].

2. Differential equations with regular singularities

2.1. Regular singularities for ordinary equations

The theory of ordinary differential equations with regular singularities has been well established for many years (cf. Hilb[a]). Since many of the phenomena that occur in the general theory for partial differential equations already appears in the much simpler special case of ordinary equations, we will start with a short summary of this theory.

The linear homogeneous first order system

$$\frac{du}{dt} = N(t)u$$

where $N(t)$ is an $n \times n$ matrix and u a column vector with n entries is said to have a regular singularity at $t=0$ if $N(t)$ is analytic in a neighborhood of 0 except for a pole of order 1 at $t=0$. Then it can be brought to the form

$$t\frac{du}{dt} = M(t)u \tag{2.1}$$

with $M(t)$ analytic in a neighborhood of 0.

Let $M_o = M(0)$. The following theorem can be found in almost every book on the subject, e.g., Coddington and Levinson [a] or Wasow [a].

Theorem 2.1.1 *If M_o has no eigenvalues that differ from each other by non-zero integers, then there exists a matrix function $U(t)$ with $U(0) = I$ and with analytic coefficients in a neighborhood of 0 such that by the transformation*

$$u = Uv$$

the equation (2.1) is reduced to

$$t\frac{dv}{dt} = M_o v \quad . \tag{2.2}$$

Proof: The claim is that there exists a matrix $U(t)$ such that

$$\left[t\frac{d}{dt} - M\right]U = U\left[t\frac{d}{dt} - M_o\right] \quad .$$

To find such a matrix one has to solve the following equation

$$t\frac{dU}{dt} - [M_o,U] = (M-M_o)U \quad . \tag{2.3}$$

This is done formally by putting $M = \sum_{j=0}^{\infty} M_j t^j$ and $U = \sum_{j=0}^{\infty} U_j t^j$.
The coefficients to t^j in (2.3) give

$$[U_o,M_o] = 0 \quad \text{if} \quad j=0$$

and

$$jU_j - [M_o,U_j] = \sum_{k=0}^{j-1} M_{j-k}\, U_k \quad \text{for} \quad j > 0 \; .$$

The first is solved by $U_o = I$, and the second can be solved recursively since the eigenvalues of the operator $[M_o,\cdot]$ on n^2-space are precisely the differences of the eigenvalues of M_o . Finally, one has to prove that the sum $\sum_{j=0}^{\infty} U_j t^j$ converges for small t - see Wasow [a], Theorem 5.4. □

If M_o has n distinct eigenvalues, say $\lambda_1,\ldots,\lambda_n$, eq. (2.2) can be diagonalized to

$$t\frac{d}{dt} w_k = \lambda_k w_k \qquad (k=1,\ldots,n) \; . \tag{2.4}$$

Therefore we now consider the simple equation

$$t\frac{d}{dt} w = \lambda w \quad . \tag{2.5}$$

On $\{t > 0\}$ the only analytic solutions to (2.5) are of course $w = c\cdot t^{\lambda}$ $(c \in \mathbb{C})$. Since the equation is elliptic on $\{t > 0\}$ the uniqueness even applies to hyperfunction solutions in this domain. Which hyperfunction solutions are there around 0? The answer to this question is provided by:

Lemma 2.1.2 The equation (2.5) has in any neighborhood of $t = 0$ exactly two linearly independent hyperfunction solutions w for all $\lambda \in \mathbb{C}$. They are given as follows:

a) <u>If</u> $\lambda \notin -\mathbb{N}$ <u>by</u> t_+^λ <u>and</u> t_-^λ .

b) <u>If</u> $\lambda \in -\mathbb{N}$ <u>by</u> $\delta^{(-\lambda-1)}$ <u>and</u> P.V. $t^\lambda = \left[\frac{1}{2}\frac{\operatorname{Im} z}{|\operatorname{Im} z|} z^\lambda\right]$.

<u>Proof</u>: It is easily seen that the functions mentioned are solutions to (2.5). Therefore it is enough to prove that the stalk at 0 of the sheaf of solutions has dimension at most 2.

Let $f \in \mathcal{B}(V)$ be a solution defined in a neighborhood V of 0, and choose a defining function $F \in \mathcal{O}(W\setminus V)$ for f on some complex neighborhood W of V. Then $(t\frac{d}{dt} - \lambda)f = 0$ means that the function $(z\frac{d}{dz} - \lambda)F$ on $W\setminus V$ extends holomorphically to W. Hence in a neighborhood of 0 we can expand it in a power series:

$$(z\frac{d}{dz} - \lambda)F = \Sigma_{n=0}^{\infty} a_n z^n \ .$$

Define a function G, holomorphic in a neighborhood of 0, by

$$G(z) = \Sigma_{n=0,\, n\neq\lambda}^{\infty} a_n (n-\lambda)^{-1} z^n \ .$$

Put $F' = F - G$, then $f = [F']$ holds in a neighborhood of 0.

Assume first $\lambda \notin \mathbb{Z}_+$. Then we see that

$$(2.6) \qquad (z\frac{d}{dz} - \lambda)F' = 0$$

and thus every solution has a defining function F' on a neighborhood of 0 in $\mathbb{C}\setminus\mathbb{R}$ satisfying (2.6). However, since the equation (2.6) is elliptic on $\mathbb{C}\setminus\mathbb{R}$ the dimension of the space of such functions is 2 (the number of components of $\mathbb{C}\setminus\mathbb{R}$). This proves the assertion in this case.

If $\lambda \in \mathbb{Z}_+$ then a_λ is independent of the choice of defining function F, and thus $\varphi(f) = a_\lambda$ defines a linear form on the solution space. We claim that the kernel of φ has dimension ≤ 1. In fact, if $a_\lambda = 0$ it again follows that (2.6) holds. This time, however, the solution $F'(z) = z^\lambda$ to (2.6) gives $[F'] = 0$, and therefore $\dim \ker \varphi \leq 1$. □

Notice that all the solutions to (2.5) are distributions.

Thus, by the indicated method, the system of equations (2.1) is completely solved when the eigenvalues of $M(0)$ are all distinct and do not differ from each other by integers. We do not need it here, but there are methods for dealing with the general equation (cf. Wasow [a], Section 17).

Finally, we consider higher order equations. The equation

$$\frac{d^n f}{dt^n} + p_1(t)\frac{d^{n-1}f}{dt^{n-1}} + \ldots + p_n(t)f = 0$$

is said to have a <u>regular singularity</u> at 0 if the coefficient functions $p_j(t)$ are analytic in a neighborhood of 0 except for poles at 0 of order at most j $(j = 1,\ldots,n)$.

Then it can be brought to the form

$$\left[\left(t\frac{d}{dt}\right)^n + a_1(t)\left(t\frac{d}{dt}\right)^{n-1} + \ldots + a_n(t)\right] f = 0 \tag{2.7}$$

with a_j analytic in a neighborhood of 0.

Let $u_j = (t\frac{d}{dt})^{j-1} f$ for $j = 1,\ldots,n$ and let u be the column vector formed by $u_1,\ldots,u_n$. This reduces (2.7) to the system (2.1) with

$$M(t) = \begin{bmatrix} 0 & 1 & & & \\ & 0 & \ddots & & \\ & & \ddots & 1 & \\ & & & 0 & 1 \\ -a_n(t) & -a_{n-1}(t) & \cdots & -a_2(t) & -a_1(t) \end{bmatrix}.$$

The characteristic equation of $M(0)$

$$\lambda^m + a_1(0)\lambda^{n-1} + \ldots + a_n(0) = 0$$

is called the <u>indicial equation</u> of (2.7). If its roots are all distinct and do not differ from each other by integers the equation can be completely solved as we saw above. (In particular, it follows that all hyperfunction solutions are distributions, cf. also Komatsu [c], Theorem 3.5).

2.2. <u>Regular singularities for partial differential equations</u>

In the remaining of this chapter we outline the results of Kashiwara and Oshima on boundary value problems for systems of equations with regular singularities. In this section, the concept of such systems is defined.

Let M be an $n+\ell$ dimensional real analytic manifold with local coordinates $(x,t) = (x_1,\dots,x_n,t_1,\dots,t_\ell)$, and let $N_1,\dots,N_\ell$ be hypersurfaces of M such that N_j in the local coordinates is given by $t_j = 0$ $(j=1,\dots,\ell)$. Put $N = N_1 \cap \dots \cap N_\ell$.

Let $P_1,\dots,P_\ell$ be analytic differential operators on M of order $r_1,\dots,r_\ell$ respectively. Consider the system of differential equations

$$\mathcal{M} : P_j u = 0 \quad (j=1,\dots,\ell) .$$

Though the theory could be done in higher generality, it is assumed that the P_j's mutually commute.

Definition. The system $\mathcal{M}$ has regular singularities along the walls $N_1,\dots,N_\ell$ with the edge N if

(I) P_j is of the form $P_j(x,t,t\frac{\partial}{\partial x}, t\frac{\partial}{\partial t})$ where, for each point (x,t), $P_j(x,t,v,s)$ is a polynomial in $v \in \mathbb{C}^{\ell n}$ and $s \in \mathbb{C}^\ell$, and by definition $t\frac{\partial}{\partial x} = (t_i \frac{\partial}{\partial x_j})_{i=1,\dots,\ell;\, j=1,\dots,n}$, $t\frac{\partial}{\partial t} = (t_i \frac{\partial}{\partial t_i})_{i=1,\dots,\ell}$.

(II) The degree of $a_j(x,s) = P_j(x,0,0,s)$ is r_j for all x, and for each x only $s=0$ solves

$$a_1^0(x,s) = \dots = a_n^0(x,s) = 0 ,$$

where $a_j^0(x,s)$ is the homogeneous part of degree r_j of $a_j(x,s)$.

The polynomial $a_j(x,s)$ is called the indicial polynomial for P_j and the roots $s(x) \in \mathbb{C}^\ell$ to the system of equations

$$a_1(x,s) = \dots = a_\ell(x,s) = 0 \tag{2.8}$$

are called the characteristic exponents of $\mathcal{M}$.

Using Bezout's theorem (Shafarewich [a] p. 198), it follows from (II) that counted with multiplicities there are exactly $r = \prod_{j=1}^{\ell} r_j$ characteristic exponents for each x. We denote these by $s_\nu(x) \in \mathbb{C}^\ell$ $(\nu = 1,\dots,r)$ and write $s_\nu(x) = (s_{\nu,1}(x),\dots,s_{\nu,\ell}(x))$. In most applications the characteristic exponents do not depend on x, or at most only depend on some coordinates x_i for which $\frac{\partial}{\partial x_i}$ does not enter into the expressions for $P_1,\dots,P_\ell$.

Note that the definitions of regular singularities and indicial polynomials are invariant under (analytic) changes of coordinates if we demand that $N_j = \{t_j' = 0\}$ in the new coordinates (x', t') also.

There is a concept slightly weaker than regular singularities:

Definition. The system $\mathcal{M}$ has regular singularities in the weak sense along the walls $N_1, \ldots, N_\ell$ with the edge N if there is a positive integer m such that (I) and (II) hold for the system obtained by substituting $t_j = (t_j')^m$ $(j = 1, \ldots, \ell)$.

(Notice that the "change of coordinates" given by $t_j = (t_j')^m$ is not analytic).

In other words, P_j is of the form

$$P_j = A_j\left(x, t, t\frac{\partial}{\partial t}\right) + \sum_{i=1}^{\ell} t_i Q_{ij}\left(t, x, \frac{\partial}{\partial x}, t\frac{\partial}{\partial t}\right)$$

and $a_j(x,s) = A_j(x,o,s)$ satisfies (II). Again $a_j(x,s)$ is called the indicial polynomial and the roots to (2.8) the characteristic exponents.

2.3. Boundary values for a single equation

The problem of determining Cauchy data ("boundary values") for a hyperfunction solution u of an analytic differential equation on non-characteristic surfaces has been studied by Komatsu [a] and Schapira [b]. For details, see Hörmander [b], Corollary 9.5.4. The assignment of the boundary values to u is done as follows: Suppose u is defined on $\{t > 0\}$ where $t = 0$ is non-characteristic for the r'th order equation $Pu = 0$. Then u is uniquely extended to a hyperfunction $\tilde{u}$, defined on the whole space but with support on $\{t \geq 0\}$, satisfying

$$(2.9) \qquad P\tilde{u} = \sum_{\nu=1}^{r} \varphi_j \otimes \delta^{(\nu-1)}(t)$$

for some unique hyperfunctions $\varphi_1, \ldots, \varphi_r$ on $\{t = 0\}$. $\varphi_1, \ldots, \varphi_r$ are called the boundary values of u. The generalization to the equation with regular singularities by Kashiwara and Oshima takes a similar course, reflected in the two theorems below.

Let P be an analytic differential operator with regular singularities in the weak sense along N, of order r and with the characteristic exponents $s_1,\dots,s_r$. Let M_+ be given by $t > 0$ in the local coordinates. Suppose $u \in \mathcal{B}(M_+)$ satisfies $Pu = 0$.

Theorem 2.3.1 (Kashiwara and Oshima [a], Corollary 4.7) Assume $s_\nu(x) \notin -\mathbb{N}$ for all $x \in N$ and $\nu = 1,\dots,r$. Then there exists a unique hyperfunction $\tilde{u} \in \mathcal{B}(M)$ which satisfies:

(i) $\tilde{u}|_{M_+} = u$,

(ii) $\operatorname{supp} \tilde{u} \subset \overline{M}_+$,

(iii) $P\tilde{u} = 0$.

Notice that the assumption on s_ν is important. As a matter of fact, Lemma 2.1.2 b shows that if this assumption does not hold, there can exist solutions on M_+ which cannot be extended in the way prescribed by Theorem 2.3.1.

Let X be a complex neighborhood of M and put $\Lambda = \{((x,0), dt\infty) \in P^*X\}$ and $\Lambda^+ = \{((x,0), \sqrt{-1}\, dt\infty) \in \sqrt{-1}\, S^*M\}$.

Theorem 2.3.2 (Kashiwara and Oshima [a], Theorem 3.1.4 and p. 175) Assume that $s_\nu(x) - s_{\nu'}(x) \notin \mathbb{Z}$ for all $\nu \neq \nu'$, $x \in N$. Then, in a neighborhood of Λ, there exist microdifferential operators $A_\nu(x, D_x, D_t)$ $(\nu = 1,\dots,r)$ of order 0 with principal symbol equal to 1 on Λ such that the transformation

$$u = \Sigma_{\nu=1}^{r} A_\nu(x, D_x, D_t) v_\nu \tag{2.10}$$

gives an isomorphism between the systems

$$\mathcal{M} : Pu = 0$$

and

$$\mathcal{N} : \left(t \frac{\partial}{\partial t} - s_\nu(x)\right) v_\nu = 0 \quad (\nu = 1,\dots,r)$$

Here u and v_ν are microfunctions defined near Λ^+. The isomorphism of the form (2.10) between $\mathcal{M}$ and $\mathcal{N}$ is unique.

The proofs of these two theorems (and the generalization of Theorem 2.3.2 to systems of differential equations - cf. Section 2.5) take up the major part of Kashiwara and Oshima [a] and rely on deep results from Sato et al. [a]. There is not enough space (nor author's insight) to go into these proofs here. In the following section, however, we will prove these theorems (except for the uniqueness statement of Theorem 2.3.2) in a special case (which in fact was the motivating example for Kashiwara and Oshima). The proofs we shall give follow the lines of Kashiwara's and Oshima's general proof.

Observe that eq. (2.9) can be derived from Theorem 2.3.1 as follows: If P is elliptic at $t = 0$, then $t^r P$ has regular singularities there with characteristic exponents $0,1,\dots,r-1$. From Theorem 2.3.1 we have the extension $\tilde{u}$ with support in $\overline{M^+}$ satisfying $t^r P\tilde{u} = 0$, which is equivalent to (2.9).

We will now show how to apply these theorems to define boundary values on N of the solutions u on M^+ to $Pu = 0$.

First we assume that P has regular singularities and satisfies both $s_\nu(x) \notin -\mathbb{N}$ and $s_\nu(x) - s_{\nu'}(x) \notin \mathbb{Z}$ for all $\nu \neq \nu'$ and $x \in N$. Then if $u \in \mathcal{B}(M_+)$ solves $Pu = 0$ we get $\tilde{u}$ from Theorem 2.3.1, and we can apply Theorem 2.3.2 to the microfunction $\mathrm{sp}\,\tilde{u}$.

Since the system $\mathcal{N}$ is so simple, we can solve it completely (cf. Lemma 2.1.2) and get that $v_\nu(x,t) = \mathrm{sp}[\varphi_\nu(x) \otimes t_+^{s_\nu(x)}]$ for a hyperfunction $\varphi_\nu \in \mathcal{B}(N)$ uniquely determined by v_ν. So, in a unique fashion, we get

$$\mathrm{sp}\,\tilde{u} = \Sigma_{\nu=1}^{r} A_\nu(x,D_x,D_t)\mathrm{sp}[\varphi_\nu(x) \otimes t_+^{s_\nu(x)}] \tag{2.11}$$

and define that φ_ν is the <u>boundary value</u> of u on N with respect to the characteristic exponent s_ν $(\nu = 1,\dots,r)$.

Note that if $\varphi_1 = \dots = \varphi_r = 0$ then $\mathrm{sp}\,\tilde{u} = 0$ by (2.11) whence $u = 0$ in a neighborhood of N by Proposition 1.5.7. Thus u is uniquely determined in a neighborhood of N by its boundary values on N.

Weakening the conditions on P, assume that P has regular singularities in the weak sense and that $s_\nu(x) - s_{\nu'}(x) \notin \mathbb{Z}$ for $\nu \neq \nu'$. Then the boundary values of u can still be defined. Put $u'(x,t') = t^k u(x,t)$ where $t = (t')^m$, and $P'(x,t',\frac{\partial}{\partial x},\frac{\partial}{\partial t'}) = t^k P(x,t,\frac{\partial}{\partial x},\frac{\partial}{\partial t})t^{-k}$. Then $P'u' = 0$ and P' has regular singularities along N with the characteristic exponents $k+s_\nu$ $(\nu = 1,\dots,r)$. For k sufficiently big the boundary values of u' are defined by the procedure above (m is given by the definition in Section 2.2). By Kashiwara and Oshima [a], Lemma 5.13, the boundary values are independent of the choice of k. They are then called the boundary values of u.

2.4. Example

Let $M = \mathbb{R}^2$ with coordinates (x,t) . For $s \in \mathbb{C}$ let

$$P_s = t^2\left(\frac{\partial^2}{\partial t^2} + \frac{\partial^2}{\partial x^2}\right) - s(s-1) \quad . \tag{2.12}$$

Then P_s has regular singularities along the x-axis with the characteristic exponents s and $1-s$. We will consider Theorems 2.3.1 and 2.3.2 for this particular operator.

Let $X = \mathbb{C}^2$ and use the coordinates (y,z) on X . Let $Y \subset X$ be the submanifold given by $z = 0$. As in Example 1.3.9 we consider the sheaf $\mathcal{H}^1_Y(\mathcal{O})$ on Y . When $V \subset Y$ is an open subset the space of sections of $\mathcal{H}^1_Y(\mathcal{O})$ over V is

$$\mathcal{O}(U \setminus V)/\mathcal{O}(U) \tag{2.13}$$

where $U = \{(y,z) \in X \mid (y,0) \in V\}$.

Denote by P_s also the extension of (2.11) to a holomorphic differential operator on X :

$$P_s = z^2\left(\frac{\partial^2}{\partial z^2} + \frac{\partial^2}{\partial y^2}\right) - s(s-1) \quad .$$

Since P_s operates on $\mathcal{O}$ it induces an action on $\mathcal{H}^1_Y(\mathcal{O})$. The main step in the proof of Theorem 2.3.1 consists of proving the following theorem, for which we will give an elementary proof in the case at hand.

Theorem 2.4.1 If $s, 1-s \notin -\mathbb{N}$ then the endomorphism $P_s : \mathcal{H}^1_Y(\mathcal{O}) \longrightarrow \mathcal{H}^1_Y(\mathcal{O})$ is bijective.

For the proof we need an elementary lemma, the proof of which we leave to the reader.

For $x \in \mathbb{C}$ and $k \in \mathbb{N}$ define

$$(x,k) = \prod_{\ell=0}^{k-1} (x+\ell) \tag{2.14}$$

and also $(x,0) = 1$.

Lemma 2.4.2 Fix $x \in \mathbb{C} \setminus \frac{1}{2}\mathbb{Z}$. There exists a constant $a_x > 0$ such that for all $m \in \frac{1}{2}\mathbb{Z}$ and $k \in \mathbb{N}$:

$$|(x-m,k)| \geq a_x \, 2^{-2k}(k-1)!$$

<u>This estimate also holds for</u> $x \in \frac{1}{2}\mathbb{Z}$ <u>if we require</u> $m < x$.

<u>Proof of Theorem 2.4.1</u>: Let V be an open ball in Y around 0 of radius, say, $r > 0$. We want to prove that the endomorphism of the space (2.13) induced by P_s is bijective. Because of the invariance of P_s under translations of y this proves that P_s is locally bijective, from which the theorem follows.

Using a Laurent series expansion in z and a power series expansion in y we identify $\mathcal{O}(U \setminus V)$ with the space

$$\{\Sigma_{i=0}^{\infty} \Sigma_{j=-\infty}^{\infty} a_{ij}\, y^i z^{-j} \mid \forall \rho < r \ \forall \varepsilon > 0 \ \exists C > 0 \ \forall i,j : |a_{ij}| \leq C\rho^{-i}\varepsilon^{j}\} .$$

Then $\mathcal{O}(U)$ corresponds to the subspace given by $a_{ij} = 0$ for $j > 0$. Therefore $\mathcal{O}(U \setminus V) / \mathcal{O}(U)$ is identified with the space

$$A = \{(a_{ij})_{i \geq 0,\, j > 0} \mid \forall \rho < r \ \forall \varepsilon > 0 \ \exists C > 0 \ \forall i,j : |a_{ij}| \leq C\rho^{-i}\varepsilon^{j}\} .$$

The action of P_s on A is easily seen to be given by $P_s a = b$ where $b \in A$ is identified by

$$(2.15) \qquad b_{ij} = (j(j+1) - s(s-1))a_{ij} + (i+2)(i+1)a_{i+2,\, j+2} .$$

The claim is that $a \mapsto b$ is bijective on A . Thus Theorem 2.4.1 is reduced to some estimates on power series.

Define for i, j , and k nonnegative integers a complex number $c(i,j,k)$ by

$$c(i,j,k) = \frac{(-1)^k (i+1,\, 2k)}{2^{2k+2}(\frac{s}{2}+\frac{j}{2},\, k+1)(\frac{1-s}{2}+\frac{j}{2},\, k+1)} .$$

Then we see from Lemma 2.4.2 that

$$|c(i,j,k)| \leq C \, \frac{2^{2k}(i+1,\, 2k)}{(k!)^2} = C2^{2k} \binom{i+2k}{2k}\binom{2k}{k}$$

for all i, j , and k for some constant C only depending on s (here we have used that $s, 1-s \notin -\mathbb{N}$). Using that the binomial coefficient $\binom{n}{m}$ is smaller than 2^n for all $m \leq n$, we get the following estimate

$$(2.16)\qquad |c(i,j,k)| \le C2^{i+6k} .$$

Now let $a \in A$ and define $b \in A$ by (2.14). By induction one proves for $n = 1,2,\ldots$ that

$$a_{ij} = \sum_{k=0}^{n-1} c(i,j,k) b_{i+2k,\, j+2k} - (i+2n-1)(i+2n) c(i,j,n-1) a_{i+2n,\, j+2n} .$$

Then (2.16) together with the growth condition on a_{ij} implies that we have

$$(2.17)\qquad a_{ij} = \sum_{k=0}^{\infty} c(i,j,k) b_{i+2k,\, j+2k} .$$

Thus $a \to b$ is injective.

Conversely, let $b \in A$. Then the growth condition on b_{ij} together with the estimate (2.16) ensure that the series (2.17) converges for all i, j and defines an element a of A. Inserting (2.17) into (2.15) then gives $b = P_s a$. □

Let $M_o = \{(t,x) \in M \mid t = 0\}$ and let

$$\mathcal{B}_{M_o}(M) = \{f \in \mathcal{B}(M) \mid \operatorname{supp} f \subset M_o\} .$$

<u>Corollary 2.4.3</u> <u>If</u> $s, 1-s \notin -\mathbb{N}$ <u>then</u>

$$P_s : \mathcal{B}_{M_o}(M) \to \mathcal{B}_{M_o}(M)$$

<u>is bijective</u>.

<u>Proof</u>: The point is of course to relate the space $\mathcal{B}_{M_o}(M)$ to the sheaf $\mathcal{H}^1_Y(\mathcal{O})$.

Let $\mathcal{B}_M = \mathcal{H}^2_M(\mathcal{O})$ denote the sheaf of hyperfunctions on M, then we can identify the space $\mathcal{B}_{M_o}(M)$ with the zero'th local cohomology space of $\mathcal{B}_M$ on M_o:

$$\mathcal{B}_{M_o}(M) \cong H^0_{M_o}(M; \mathcal{H}^2_M(\mathcal{O})) .$$

Using Komatsu [d], Theorem 1.9, we can identify this space with the second local cohomology space of $\mathcal{O}$ on M_o:

$$H^0_{M_o}(M; \mathcal{H}^2_M(\mathcal{O})) \cong H^2_{M_o}(X; \mathcal{O}) .$$

On the other hand, using loc. cit. once more, we have

$$H^2_{M_o}(X;\mathcal{O}) \cong H^1_{M_o}(Y;\mathcal{H}^1_Y(\mathcal{O}))$$

and since P_s is bijective on the sheaf $\mathcal{H}^1_Y(\mathcal{O})$ it is also bijective on its local cohomology space. □

Let $M^+ = \mathbb{R} \times]0,\infty[$ and $\overline{M^+} = \mathbb{R} \times [0,\infty[$.

<u>Corollary 2.4.4</u> <u>If</u> $s, 1-s \notin -\mathbb{N}$ <u>then every hyperfunction solution</u> $f \in \mathcal{B}(M^+)$ <u>to the equation</u> $P_s f = 0$ <u>has a unique extension to a hyperfunction</u> $\tilde{f} \in \mathcal{B}(\mathbb{R}^2)$ <u>satisfying</u>:

$$\tilde{f}|_{M^+} = f \ , \qquad \operatorname{supp} \tilde{f} \subset \overline{M^+}, \ \underline{\text{and}} \ \ P_s\tilde{f} = 0 \ .$$

<u>Proof</u>: Since $\mathcal{B}$ is flabby there exists a hyperfunction $g \in \mathcal{B}(\mathbb{R}^2)$ with $\operatorname{supp} g \subset \overline{M^+}$ and $g|_{M^+} = f$. Then $\operatorname{supp} P_s g \subset M_o$ and by the preceding corollary there is a unique hyperfunction $h \in \mathcal{B}(\mathbb{R}^2)$ with support on M_o such that $P_s h = P_s g$. Then $\tilde{f} = g - h$ satisfies the stated conditions, and is unique. □

Now we will show how the isomorphism (2.10) is constructed for the operator P_s .

Suppose u is a microfunction solution to $\mathcal{M}$: $P_s u = 0$ in a neighborhood of $(x,0,\sqrt{-1}\, dt\, \infty) \in \sqrt{-1}\, S^*M$. As with the ordinary equation (2.7) we transform $\mathcal{M}$ into a first order system by setting

$$(2.18) \qquad w_1 = u \quad \text{and} \quad w_2 = t\frac{\partial}{\partial t} u \ .$$

Let w be the column vector with entries w_1 and w_2 . Then we have a first order system

$$t\frac{\partial}{\partial t} w = \begin{pmatrix} 0 & 1 \\ -t^2 \dfrac{\partial^2}{\partial x^2} + s(s-1) & 1 \end{pmatrix} w$$

but actually we prefer the elements of the matrix on the right hand side to be of order ≤ 0 . This can easily be arranged as follows:

Let R be the micro-differential operator

$$R = \frac{\frac{\partial^2}{\partial x^2}}{\frac{\partial^2}{\partial t^2} + \frac{\partial^2}{\partial x^2}} .$$

It then follows that

$$t \frac{\partial}{\partial t} w_2 = \left(t \frac{\partial}{\partial t}\right)^2 u = \left\{-t^2 \frac{\partial^2}{\partial x^2} + t \frac{\partial}{\partial t} + s(s-1)\right\} u$$

$$= \left\{-R\left(\frac{\partial^2}{\partial t^2} + \frac{\partial^2}{\partial x^2}\right) t^2 + t \frac{\partial}{\partial t} + s(s-1)\right\} u$$

$$= \left\{-R\left[t^2\left(\frac{\partial^2}{\partial t^2} + \frac{\partial^2}{\partial x^2}\right) + 4t \frac{\partial}{\partial t} + 2\right] + t \frac{\partial}{\partial t} + s(s-1)\right\} u$$

$$= (s(s-1)-(2+s(s-1))R)w_1 + (1-4R)w_2 .$$

Thus

$$t \frac{\partial}{\partial t} w = Lw$$

where

$$L = \begin{pmatrix} 0 & 1 \\ s(s-1)-(2+s(s-1))R & 1-4R \end{pmatrix} .$$

Note that L takes the form

$$L_o = \begin{pmatrix} 0 & 1 \\ s(s-1) & 1 \end{pmatrix}$$

at $((x,0),dt\infty)$, and that the eigenvalues of L_o are s and $1-s$.

Now the operator $t \frac{\partial}{\partial t} - L$ is diagonalized by a recursive procedure very similar to the proof of Theorem 2.1.1, as follows:

<u>Theorem 2.4.5</u> <u>There exists an invertible</u> 2×2 <u>matrix</u> U <u>of micro-differential operators of order</u> 0 <u>in a neighborhood of</u> $((x,0),dt\infty)$ <u>such that</u>

$$U^{-1}\left(t \frac{\partial}{\partial t} - L\right) U = t \frac{\partial}{\partial t} - \tilde{L}_o \tag{2.19}$$

<u>where</u> $\tilde{L}_o = \begin{pmatrix} s & 0 \\ 0 & 1-s \end{pmatrix}$.

Proof: First we apply to L the inner automorphism from $GL(2,\mathbb{C})$ which diagonalizes L_o,

$$\tilde{L}_o = \frac{1}{2s-1}\begin{pmatrix} s-1 & 1 \\ s & 1 \end{pmatrix} L_o \begin{pmatrix} 1 & -1 \\ s & s-1 \end{pmatrix} = \begin{pmatrix} s & 0 \\ 0 & 1-s \end{pmatrix}$$

and get

$$\tilde{L} = \frac{1}{2s-1}\begin{pmatrix} s-1 & 1 \\ s & 1 \end{pmatrix} L \begin{pmatrix} 1 & -1 \\ s & s-1 \end{pmatrix} = \tilde{L}_o + BR$$

where we define the complex matrix B by

$$B = \begin{pmatrix} b(s) & b(1-s) \\ b(s) & b(1-s) \end{pmatrix}$$

with $b(r) = \dfrac{(1+r)(2+r)}{1-2r}$ $(r \neq \frac{1}{2})$. Put $U = \begin{pmatrix} 1 & -1 \\ s & s-1 \end{pmatrix} \tilde{U}$, then (2.19) is equivalent to

$$\tilde{U}^{-1}\left(t\frac{\partial}{\partial t} - \tilde{L}\right)\tilde{U} = t\frac{\partial}{\partial t} - \tilde{L}_o \;. \tag{2.20}$$

Since $\tilde{L}$ does not depend on (x,t) and is homogeneous of degree 0 in the cotangent variable, it is reasonable that the same holds for $\tilde{U}$. Therefore, we seek a 2×2 matrix $A = A(z)$ of functions $a_{ij}(z)$ analytic in a neighborhood of $z = 0 \in \mathbb{C}$, such that (2.20) holds with

$$\tilde{U} = A\left(\frac{\partial}{\partial x}\left(\frac{\partial}{\partial t}\right)^{-1}\right)$$

(cf. Proposition 1.6.3).

Since $\left[\dfrac{\partial^k}{\partial t^k}, t\dfrac{\partial}{\partial t}\right] = k\dfrac{\partial^k}{\partial t^k}$ for all $k \in \mathbb{Z}$ we get that $\left[t\dfrac{\partial}{\partial t}, \tilde{U}\right] = \left(z\dfrac{dA}{dz}\right)\left(\dfrac{\partial}{\partial x}\left(\dfrac{\partial}{\partial t}\right)^{-1}\right)$, using the Taylor series of $A(z)$.

Therefore, multiplying (2.20) from the left with $\tilde{U}$, we get the following equation of 2×2 matrices of microdifferential operators:

$$\left(z\frac{dA}{dz}\right)\left(\frac{\partial}{\partial x}\left(\frac{\partial}{\partial t}\right)^{-1}\right) = \tilde{L}A\left(\frac{\partial}{\partial x}\left(\frac{\partial}{\partial t}\right)^{-1}\right) - A\left(\frac{\partial}{\partial x}\left(\frac{\partial}{\partial t}\right)^{-1}\right)\tilde{L}_o.$$

Replacing $\dfrac{\partial}{\partial x}\left(\dfrac{\partial}{\partial t}\right)^{-1}$ with z we get an equation of 2×2 matrices of analytic functions, which we in analogy with equation (2.3) write as

$$z\frac{dA}{dz} - [\tilde{L}_o, A] = (\tilde{L} - \tilde{L}_o)A \;. \tag{2.21}$$

Here, $\widetilde{L}$ is the matrix given by

$$\widetilde{L} = \widetilde{L}_o + B\,\frac{z^2}{1+z^2} .$$

We now expand A in a power series of matrices $A = \Sigma_{j=0}^{\infty} A_j z^j$, insert this into (2.21), and get that

$$\Sigma_{j=0}^{\infty}(j\,A_j - [\widetilde{L}_o, A_j])z^j = \Sigma_{j=0}^{\infty} B\,\frac{z^2}{1+z^2}\,A_j z^j .$$

Multiplying by $1+z^2$, this leads to the following relations for A_j :

$$\text{(2.22)}\qquad \begin{cases} [\widetilde{L}_o, A_0] = A_1 - [\widetilde{L}_o, A_1] = 0 \\[2mm] jA_j - [\widetilde{L}_o, A_j] = BA_{j-2} - (j-2)A_{j-2} + [\widetilde{L}_o, A_{j-2}] \quad (j \geq 2) \end{cases}$$

which can be solved recursively since $2s - 1 \notin \mathbb{Z}$.

Taking $A_0 = 1$ and $A_1 = 0$ it is straightforward but tiresome to check that $A_{2j-1} = 0$ and

$$A_{2j} = (-1)^j \begin{bmatrix} \dfrac{\left(\frac{s}{2}+\frac{1}{2}, j\right)\left(\frac{s}{2}+1, j\right)}{j!\,(s-\frac{1}{2}, j)} & \dfrac{b(1-s)}{2s-3}\,\dfrac{\left(-\frac{s}{2}+2, j-1\right)\left(-\frac{s}{2}+\frac{5}{2}, j-1\right)}{(j-1)!\,(-s+\frac{5}{2}, j-1)} \\[6mm] -\dfrac{b(s)}{2s+1}\,\dfrac{\left(\frac{s}{2}+\frac{3}{2}, j-1\right)\left(\frac{s}{2}+2, j-1\right)}{(j-1)!\,(s+\frac{3}{2}, j-1)} & \dfrac{\left(-\frac{s}{2}+1, j\right)\left(-\frac{s}{2}+\frac{3}{2}, j\right)}{j!\,(-s+\frac{1}{2}, j)} \end{bmatrix}$$

solves (2.22). We have used the notation of (2.14) for (x,k) .

The convergence of $\sum_{j=0}^{\infty} A_j z^j$ for small z is now immediate, in fact

$$A(z) = \begin{bmatrix} F\left(\frac{s}{2}+\frac{1}{2}, \frac{s}{2}+1, s-\frac{1}{2}, -z^2\right) & -\dfrac{b(1-s)}{2s-3}\, z^2 F\left(-\frac{s}{2}+2, -\frac{s}{2}+\frac{5}{2}, -s+\frac{5}{2}, -z^2\right) \\[6mm] \dfrac{b(s)}{2s+1}\, z^2 F\left(\frac{s}{2}+\frac{3}{2}, \frac{s}{2}+2, s+\frac{3}{2}, -z^2\right) & F\left(-\frac{s}{2}+1, -\frac{s}{2}+\frac{3}{2}, -s+\frac{1}{2}, -z^2\right) \end{bmatrix}$$

where F denotes the standard hypergeometric function (cf. Erdélyi et al. [a]).

This completes the proof of Theorem 2.4.5. □

Writing the elements in the first row of U as A_1 and A_2 we see by (2.18) and (2.19) that the existence of the isomorphism (2.10) is now established.

From the proof of Theorem 2.4.5 we even get formulas for A_1 and A_2. Since

$$U = \begin{pmatrix} 1 & -1 \\ s & s-1 \end{pmatrix} \tilde{U}$$

we get $A_1 = Q_s$ and $A_2 = -Q_{1-s}$ where

$$Q_s = F\left(\frac{s}{2}+\frac{1}{2}, \frac{s}{2}+1, s-\frac{1}{2}, -z^2\right) - \frac{b(s)}{2s+1} z^2 F\left(\frac{s}{2}+\frac{3}{2}, \frac{s}{2}+2, s+\frac{3}{2}, -z^2\right)$$

$$= F\left(\frac{s}{2}+\frac{1}{2}, \frac{s}{2}+1, s+\frac{1}{2}, -z^2\right)$$

with $z = \frac{\partial}{\partial x}\left(\frac{\partial}{\partial t}\right)^{-1}$.

2.5. Boundary values for a system of equations

Let $\mathcal{M}$ be a system of differential equations with regular singularities in the weak sense along the walls $N_1, \ldots, N_\ell$ with the edge N, and suppose u is a solution to $\mathcal{M}$ on $M_+ = \{t_j > 0, j = 1, \ldots, \ell\}$. The boundary values of u on N are then defined under the following conditions:

(A) $s_\nu(x) - s_{\nu'}(x) \notin \mathbf{Z}^\ell$ for $\nu \neq \nu'$, $x \in N$.

(B) For each j there is a differential operator $Q_j = Q_j(x, t, \frac{\partial}{\partial x}, t\frac{\partial}{\partial t})$ in a neighborhood of N, such that $Q_j u = 0$ and Q_j has regular singularities in the weak sense along N_j.

Assume (A) and (B). Then the boundary values are defined as follows. By the argument on p. 41, it can be assumed that $\mathcal{M}$ has regular singularities, and that no characteristic exponents of Q_j $(j = 1, \ldots, \ell)$ are negative integers. Using (B) and Theorem 2.3.1, u is extended to a hyperfunction $\tilde{u}$ with support on $\overline{M_+}$. By Kashiwara and Oshima [a] Corollary 5.11, $\tilde{u}$ also solves $\mathcal{M}$. Since Theorem 2.3.2 can be generalized to systems of differential equations with regular singularities (loc. cit. Thm. 5.3), this gives the analogue of (2.11):

$$\text{(2.23)} \qquad sp\,\tilde{u} = \Sigma_{\nu=1}^{r} A_\nu(x, D_x, D_t) sp[\varphi_\nu(x) \otimes \prod_{j=1}^{\ell} (t_j)_+^{s_{\nu,j}(x)}] .$$

Then by definition φ_ν is the boundary value of u on N with respect to s_ν (loc. cit. Definition 5.7).

We will now mention some important properties of this process of taking the boundary values. Throughout we assume (A) and (B).

2.5.1 By Proposition 1.5.7 u is uniquely determined in a neighborhood of N in M^+ by its boundary values on N.

2.5.2 It often happens that the coefficients of the operators $P_1, \ldots, P_\ell$ defining $\mathcal{M}$ depend holomorphically on some parameter, say $\lambda \in U$, where U is an open set in $\mathbb{C}^N$ for some $N \in \mathbb{N}$. In (2.12), for instance, P_s depends holomorphically on $s \in \mathbb{C}$. For the following considerations we assume that the characteristic exponents depend on λ, but not on x. Let $u = u(\lambda, x, t)$ be a solution on $U \times M_+$, which also depends holomorphically on λ. Considering the real and imaginary parts of λ as extra x-coordinates, it easily follows from the proof of (2.23) given in Kashiwara and Oshima [a] that also the boundary values $\varphi_\nu(\lambda, x)$ depend holomorphically on λ. For a fixed $\lambda \in U$, the hyperfunctions $\varphi_\nu(\lambda, \cdot)$ are the boundary values of $u(\lambda, \cdot)$. (See Oshima and Sekiguchi [a], Section 2.2 for details on the considerations of this paragraph).

For the remainder of this chapter, we assume for simplicity that the characteristic exponents $s_1, \ldots, s_r$ of $\mathcal{M}$ do not depend on x.

2.5.3 We have only defined the boundary values on N of the solution u on M to $\mathcal{M}$ locally, and our definition so far depends on the choice of local coordinates. Thus let (x,t) and (x',t') be local analytic coordinates on M, and denote by $\eta : \Omega \longrightarrow M$ and $\eta' : \Omega' \longrightarrow M$ the maps $(x,t) \longrightarrow M$ and $(x',t') \longrightarrow M$, respectively, where Ω and Ω' are neighborhoods of $(0,0)$. Assume that $N_j \cap \eta(\Omega) = \{\eta(x,t) \mid (x,t) \in \Omega,\ t_j = 0\}$ for $j = 1, \ldots, \ell$, and similarly for η'. Let φ_ν $(\nu = 1, \ldots, r)$ be the boundary values of u in coordinates (x,t), then φ_ν is a hyperfunction on $\{x \mid (x,0) \in \Omega\}$. Define φ'_ν similarly on $\{x' \mid (x',0) \in \Omega'\}$.

Theorem 2.5.4 The relation between φ_ν and φ'_ν is:

$$\varphi'_\nu(x') = \prod_{j=1}^{r} \left[\frac{\partial(\eta'^{-1} \circ \eta)}{\partial t}(x,0)\right]^{-s_{\nu,j}} \varphi_\nu(x) \tag{2.24}$$

when $(x',0) \in \Omega'$, $(x,0) \in \Omega$ and $\eta'(x',0) = \eta(x,0)$.

Proof: See Kashiwara and Oshima [a], Theorem 5.8. □

Let $T^*_{N_j}M$ denote the conormal bundle of N_j, considered as a line bundle over N, for $j = 1,\dots,\ell$. The local coordinates for $T^*_{N_j}M$ are then (x, dt_j). Since this is an oriented line bundle, one can introduce for $s \in \mathbb{C}$ the line bundle $T^*_{N_j}M^{\otimes s}$ on N for each $j = 1,\dots,r$. Denoting by (x, dt_j^s) the coordinates for this line bundle, the change of coordinates is given by

$$dt_j'^{\,s} = \left(\frac{\partial(\eta'^{-1} \circ \eta)}{\partial t_j}(x,0)\right)^s dt_j^s \quad . \tag{2.25}$$

(This can be taken as the definition of $T^*_{N_j}M^{\otimes s}$). As follows from Theorem 2.5.4, this gives a nice way of defining boundary values:

Define the line bundles on N for $\nu = 1,\dots,r$:

$$\mathcal{L}_\nu = \bigotimes_{j=1}^{\ell} T^*_{N_j}M^{\otimes s_{\nu,j}} \tag{2.26}$$

and denote by dt^{s_ν} the local section $dt_1^{s_{\nu,1}} \otimes\dots\otimes dt_\ell^{s_{\nu,\ell}}$ of this line bundle. Then the hyperfunction section

$$\varphi_\nu(x)dt^{s_\nu}$$

of $\mathcal{L}_\nu$ is independent of the choice of coordinates.

Let

$$\mathcal{B}(M_+;\mathcal{M}) = \left\{u \in \mathcal{B}(M_+) \,\middle|\, \begin{array}{l}\text{there is a neighborhood } U \text{ of } N \\ \text{such that } u \text{ solves } \mathcal{M} \text{ in } U \cap M_+\end{array}\right\}$$

and let $\mathcal{B}(N;\mathcal{L}_\nu)$ denote the space of hyperfunction valued sections of $\mathcal{L}_\nu$. Then the map

$$\beta : \mathcal{B}(M_+;\mathcal{M}) \longrightarrow \bigoplus_{\nu=1}^{r} \mathcal{B}(N;\mathcal{L}_\nu) \tag{2.27}$$

locally defined by $\beta(u) = \bigoplus_{\nu=1}^{r} \varphi_\nu(x)dt^{s_\nu}$ is well defined and independent of coordinates. β is called the boundary value map.

2.5.5 It is of great importance to know when the boundary values are analytic functions. Part (ii) of the following theorem is the crucial result on which the asymptotic expansions given in Chapters 5, 6 and 7 rely.

Theorem 2.5.6

(i) If $u \in \mathcal{B}(M_+, \mathcal{M})$ satisfies the following equation for $t_j > 0$ $(j = 1, \ldots, \ell)$

$$(2.28) \qquad u(x,t) = \sum_{\nu=1}^{r} \varphi_\nu(x,t) t_\nu^{s_\nu}$$

where $\varphi_\nu(x,t)$ is real analytic in a neighborhood of N and $t_\nu^{s_\nu} = \prod_{j=1}^{\ell} t_j^{s_{\nu,j}}$, then $\varphi_\nu(x) = \varphi_\nu(x,0)$ is the boundary value of u with respect to s_ν.

(ii) Conversely, if all the boundary values φ_ν $(\nu = 1, \ldots, r)$ of u are real analytic, then there exists unique real analytic functions $\varphi_\nu(x,t)$ in a neighborhood of N such that (2.28) holds. In particular, if $\varphi_\nu(x) \equiv 0$ for some ν, then $\varphi_\nu(x,t) \equiv 0$.

Proof: (i) See Kashiwara and Oshima [a], Proposition 5.14.
(ii) (For details, see Oshima and Sekiguchi [a], Proposition 2.16). By Kashiwara and Oshima [a], Theorem 5.3, $A_\nu(x, D_x, D_t)$ has the form

$$A_\nu(x, D_x, D_t) = \Sigma_{\beta \in \mathbb{Z}_+} P_{\nu,\beta}(x, D_x) D_t^{-\beta}$$

where $\mathbb{Z}_+ = \{0, 1, 2, \ldots\}$. This implies easily (cf. loc. cit. Lemma 5.2) that

$$A_\nu(x, D_x, D_t) sp(\varphi_\nu(x) t_+^{s_\nu})$$

$$= \Sigma_\beta P_{\nu,\beta}(x, D_x) \varphi_\nu(x) t^\beta \prod_{j=1}^{\ell} \prod_{k=1}^{\beta_j} (s_{\nu,j} + k)^{-1} sp\ t_+^{s_\nu}$$

$$= \varphi_\nu(x,t) sp\ t_+^{s_\nu}$$

where $\varphi_\nu(x,t)$ is holomorphic in a neighborhood of N and $\varphi_\nu(x,0) = \varphi_\nu(x)$. By (i), u and the function $\Sigma_{\nu=1}^{r} \varphi_\nu(x,t) t_\nu^{s_\nu}$ on M^+ both have the boundary values $\varphi_1, \ldots, \varphi_r$, from which (2.28) follows (cf. 2.5.1). The last assertion of (ii) also follows because

$$\varphi_\nu(x,t) = \Sigma_\beta P_{\nu,\beta}(x,D_x)\varphi_\nu(x)t^\beta \prod_{j=1}^{\ell} \prod_{k=1}^{\beta_j} (x_{\nu,j}+k)^{-1} . \quad \square$$

2.5.7 At a certain point of Section 6.3 we shall need the following theorem, which has recently been proved by T. Oshima. Suppose $\mathcal{M}'$ is another system of differential equations with regular singularities in the weak sense along the walls $N_1,\ldots,N_\ell$ with the edge N, and assume that $\mathcal{M}'$ has the same characteristic exponents $s_1,\ldots,s_r$ as $\mathcal{M}$. Let

$$\beta' : \mathcal{B}(M_+;\mathcal{M}') \longrightarrow \bigoplus_{\nu=1}^{r} (N;\mathcal{L}_\nu)$$

be the boundary value map for $\mathcal{M}'$.

Theorem 2.5.8 Let $u \in \mathcal{B}(M_+;\mathcal{M}) \cap \mathcal{B}(M_+;\mathcal{M}')$. Then

$$\beta(u) = \beta'(u) .$$

Proof: See Oshima [f], Corollary 4.7. Notice that if the boundary values of u for $\mathcal{M}$ are analytic, then the statement follows immediately from Theorem 2.5.6. $\square$

2.6 Notes

Except for Section 2.1, the material of which is classical (see for instance Coddington and Levinson [a]), the theory of this chapter is mainly due to M. Kashiwara and T. Oshima [a]. The definitions in Section 2.2 and the theorems 2.3.1, 2.3.2, 2.5.4 and 2.5.6 (i) are all taken from there. Theorem 2.5.6 (ii) is from Oshima and Sekiguchi [a].

For different approaches to partial differential equations with regular singularities we refer to Harish-Chandra [g], Casselman and Miličić [a], Wallach [c] and the appendix section of Knapp [b].

In Oshima [e] a simpler but not yet as powerful theory is presented. In Oshima [f] (from which Theorem 2.5.8 is taken) the theory of Kashiwara and Oshima [a] is generalized. For instance, the restriction on the characteristic exponents is removed.

The example treated in Section 2.4 is also considered in Section 0 of Kashiwara et al. [a] (but differently), and in particular the formula for Q_s is derived there. Lemma 2.1.2 is stated without proof in Sato [a]I p. 185 (see also Komatsu [c] p. 18).

3. Riemannian symmetric spaces and invariant differential operators – preliminaries

In this chapter we give a short summary of some notation and well known results which we need in the sequel. The material can be found, for instance, in Helgason's books [j] and [n], except for the results of Section 3.2, where we refer to Varadarajan [b].

3.1 Decomposition and integral formulas for semisimple Lie groups

Let G be a connected noncompact semisimple Lie group with finite center and K a maximal compact subgroup. Then G/K is a Riemannian symmetric space of the noncompact type.

Let $\mathfrak{g}$ and $\mathfrak{k}$ be the Lie algebras of G and K, respectively, and let $\langle , \rangle$ denote the Killing form on $\mathfrak{g}$. $\langle , \rangle$ is nondegenerate since $\mathfrak{g}$ is semisimple. Let $\mathfrak{p}$ denote the orthocomplement of $\mathfrak{k}$ in $\mathfrak{g}$, and θ the involution of $\mathfrak{g}$ having $\mathfrak{k}$ and $\mathfrak{p}$ as $+1$ and -1 eigenspaces. The symmetric bilinear form $B_\theta(X,Y) = \langle \theta X, Y \rangle$ is then strictly negative definite.

Let $\mathfrak{a}$ be a maximal abelian subspace of $\mathfrak{p}$. Since $\mathrm{ad}(\mathfrak{a})$ is a commutative algebra of linear transformations of $\mathfrak{g}$ symmetric with respect to B_θ, we have $\mathfrak{g} = \bigoplus_{\lambda \in \mathfrak{a}^*} \mathfrak{g}^\lambda$ where $\mathfrak{g}^\lambda = \{X \in \mathfrak{g} \mid [H,X] = \lambda(H)X, \forall H \in \mathfrak{a}\}$. Let Σ be the set of those nonzero $\lambda \in \mathfrak{a}^*$ such that $\mathfrak{g}^\lambda \neq 0$. Σ is a (possibly nonreduced) root system, its elements are called restricted roots. Fix a linear order in $\mathfrak{a}^*$ and let Σ^+ consist of the positive elements of Σ. Let Δ be the set of simple roots with respect to this order and let $\alpha_1, \ldots, \alpha_n$ be the elements of Δ. The rank of G/K is by definition equal to the dimension n of $\mathfrak{a}$.

For each $\lambda \in \Sigma$ let $m(\lambda) = \dim \mathfrak{g}^\lambda$ be its multiplicity. To facilitate some notations, let Σ' (resp. $\Sigma^{+\prime}$) consist of the elements of Σ (resp. Σ^+) each repeated according to its multiplicity. Define $\rho = \frac{1}{2} \sum_{\lambda \in \Sigma^{+\prime}} \lambda = \frac{1}{2} \sum_{\lambda \in \Sigma^+} m(\lambda)\lambda$. For each $\lambda \in \Sigma^{+\prime}$ choose a nonzero element X_λ of $\mathfrak{g}^\lambda$ such that the various X_λ corresponding to the same element of Σ^+ form a basis of $\mathfrak{g}^\lambda$. Put $X_{-\lambda} = \theta X_\lambda \in \mathfrak{g}^{-\lambda}$.

Let $\mathfrak{n} = \sum_{\lambda \in \Sigma^+} \mathfrak{g}^\lambda$ and $\overline{\mathfrak{n}} = \theta(\mathfrak{n})$. Let N (resp. $\overline{N}, A$) be the analytic subgroup of G with Lie algebra $\mathfrak{n}$ (resp. $\overline{\mathfrak{n}}, \mathfrak{a}$). Let M (resp. M^*) be the centralizer (resp. normalizer) of $\mathfrak{a}$ in K, and let $\mathfrak{m}$ be their Lie algebra. Let $\mathfrak{a}^+ = \{H \in \mathfrak{a} \mid \lambda(H) > 0 , \forall \lambda \in \Sigma^+\}$, $A^+ = \exp \mathfrak{a}^+$ and $\overline{A^+}$ the closure of A^+ in A. Put $P = MAN$, then P is a closed subgroup of G, called a minimal parabolic subgroup.

Let $W = M^*/M$, then W acts on $\mathfrak{a}$ and $\mathfrak{a}^*$ (via the Killing form) and is naturally identified with the Weyl group of Σ. Abusing notation, we will often identify an element $w \in W$ with one of its representatives in M^*.

There are several important decomposition theorems for G:

(3.1) $G = \exp \mathfrak{p}\, K$. The map

$$\mathfrak{p} \times K \ni (X,k) \longrightarrow \exp X\, k \in G$$

is an analytic diffeomorphism.

(3.2) $G = KAN$ (Iwasawa decomposition). The map

$$K \times A \times N \ni (k,a,n) \longrightarrow kan \in G$$

is an analytic diffeomorphism.

(3.3) $G = K\overline{A^+}K$ (Cartan decomposition). The map

$$K \times \overline{A^+} \times K \ni (k_1, a, k_2) \longrightarrow k_1 a k_2 \in G$$

has the fiber $\{(k_1 m, a, m^{-1}k_2) \mid m \in Z_K(a)\}$ above $k_1 a k_2$ (where $Z_K(a)$ denotes the centralizer of a in K). The map

$$K/M \times A^+ \times K \ni (k_1 M, a, k_2) \longrightarrow k_1 a k_2 \in G$$

is an analytic diffeomorphism onto an open dense set.

(3.4) $G = \cup_{w \in W} NwP$ - disjoint union (Bruhat decomposition). Let $w_o \in W$ denote the unique element of W such that $w_o \Sigma^+ = -\Sigma^+$. Then $Nw_oP = w_o\overline{N}P$. The map

$$\overline{N} \times M \times A \times N \ni (\overline{n}, m, a, n) \longrightarrow \overline{n}man \in G$$

is an analytic diffeomorphism onto an open dense set $\overline{N}P$ in G.

Let $H_1, \ldots, H_n$ be the basis of $\mathfrak{a}$ dual to $\alpha_1, \ldots, \alpha_n$. For any $\lambda \in \mathfrak{a}_c^*$ let $H_\lambda \in \mathfrak{a}_c$ be defined by $\lambda(H) = \langle H, H_\lambda \rangle$ for all $H \in \mathfrak{a}$. Then $\langle H_i, H_{\alpha_j} \rangle = \delta_{ij}$. We also have that $[X_\alpha, X_{-\alpha}] = \langle X_\alpha, X_{-\alpha} \rangle H_\alpha$ for $\alpha \in \Sigma$ and $X_{\pm\alpha} \in \mathfrak{g}^{\pm\alpha}$.

Let $\kappa : G \to K$ and $H : G \to \mathfrak{a}$ be the projections defined by the Iwasawa decomposition, i.e., by $g \in \kappa(g)\exp H(g)N$ for $g \in G$.

We shall also need the following integral formulas. Let dg, dk, da, dn and $d\bar{n}$ respectively be invariant measures on G, K, A, N and $\bar{N}$. Under suitable normalizations

$$(3.5) \qquad \int_G f(g)dg = \int_K \int_A \int_N f(kan)a^{2\rho}\, dk\, da\, dn$$

for $f \in C_0(G)$. Here a^λ stands for $\exp \lambda(H(a))$, when $a \in A$ and $\lambda \in \mathfrak{a}_c^*$.

Also for $f \in C(K)$ and $g \in G$:

$$(3.6) \qquad \int_K f(k)dk = \int_K f(\kappa(gk))\exp\langle -2\rho, H(gk)\rangle dk$$

and

$$(3.7) \qquad \int_K f(k)dk = \int_{\bar{N}} \int_M f(\kappa(\bar{n})m)\exp\langle -2\rho, H(\bar{n})\rangle dm\, d\bar{n} \quad .$$

On compact groups we use normalized measure, therefore (3.7) presumes the following normalization of $d\bar{n}$:

$$(3.8) \qquad \int_{\bar{N}} \exp\langle -2\rho, H(\bar{n})\rangle d\bar{n} = 1 \quad .$$

3.2 Parabolic subgroups

We shall be dealing with many other parabolic subgroups of G than the minimal one. In this section we therefore summarize some notation and simple results concerning these. For details we refer to Harish-Chandra [f]I Section 4, Varadarajan [b] Part II Section 6 or Warner [a] Section 1.2.

When dealing with parabolic subgroups it is convenient to work with Lie groups of a slightly more general nature than those considered in the previous section.

By a <u>real reductive Lie group with compact center</u> we will mean a real Lie group G with a maximal compact subgroup K and an involution θ of the Lie algebra $\mathfrak{g}$ of G satisfying the following axioms (3.9)-(3.12) (cf. Harish-Chandra [f]I, p. 105):

(3.9) $\mathfrak{g}$ is reductive.

(3.10) $Ad(g)$ is an inner automorphism of $\mathfrak{g}_c$ for each $g \in G$.

(3.11) K contains the center of G.

(3.12) The Lie algebra $\mathfrak{k}$ of K is the fixed point set of θ, and if $\mathfrak{p}$ denotes the -1 eigenspace of θ, then

$$(X,k) \longrightarrow \exp X \cdot k$$

is an analytic diffeomorphism of $\mathfrak{p} \times K$ onto G.

Let G be a real reductive Lie group with compact center. From (3.12) it follows that G has a finite number of connected components, since K meets each of them.

Let G_1 be the analytic subgroup with Lie algebra $[\mathfrak{g}, \mathfrak{g}]$, then

$$G/K \simeq G_1/K \cap G_1 \quad .$$

Almost everything which has been said in the previous section generalizes to this class of groups, and we will take over the notation of that section without further comment.

In particular, we have the Iwasawa decomposition $G = KAN$ and define $M = K^{\mathfrak{a}}$, the centralizer of $\mathfrak{a}$ in K. Then we have (cf. Harish-Chandra [f]I Lemma 11).

<u>Lemma 3.2.1</u> M <u>meets every component of</u> G.

By definition, a <u>parabolic subgroup</u> of G is a closed subgroup containing some conjugate of P. If it contains P itself, it is called a <u>standard</u> parabolic subgroup. Obviously any parabolic subgroup is conjugate to a standard one.

It turns out that the standard parabolic subgroups are parametrized by the subsets F of Δ. Before stating this precisely, we introduce the following notation:

For each $F \subset \Delta$ define:

$$\langle F\rangle = \{\lambda \in \Sigma \mid \lambda(H_j) = 0 \text{ for all } j \text{ with } \alpha_j \notin F\} = \Sigma \cap (\Sigma_{\alpha \in F} \mathbb{Z}\alpha)$$

$$\mathfrak{a}_F = \{H \in \mathfrak{a} \mid \alpha(H) = 0 \ \forall \alpha \in F\} = \Sigma_{\alpha_j \notin F} \mathbb{R} H_j$$

$$\mathfrak{a}(F) = \{X \in \mathfrak{a} \mid \langle X,H\rangle = 0 \ \forall H \in \mathfrak{a}_F\} = \Sigma_{\alpha \in F} \mathbb{R} H_\alpha$$

$$\mathfrak{a}^F = \mathfrak{a}_{\Delta \setminus F} = \Sigma_{\alpha_j \in F} \mathbb{R} H_j$$

$$\mathfrak{n}_F = \Sigma_{\lambda \in \Sigma^+ \setminus \langle F\rangle} \mathfrak{g}^\lambda \quad \text{and} \quad \overline{\mathfrak{n}}_F = \theta(\mathfrak{n}_F)$$

$$\mathfrak{n}(F) = \Sigma_{\lambda \in \Sigma^+ \cap \langle F\rangle} \mathfrak{g}^\lambda \quad \text{and} \quad \overline{\mathfrak{n}}(F) = \theta(\mathfrak{n}(F))$$

$$\mathfrak{m}_F = \overline{\mathfrak{n}}(F) + \mathfrak{m} + \mathfrak{a}(F) + \mathfrak{n}(F)$$

$$\mathfrak{p}_F = \mathfrak{m}_F + \mathfrak{a}_F + \mathfrak{n}_F$$

$$\mathfrak{b}_F = (\mathfrak{m}_F \cap \mathfrak{h}) + \mathfrak{a}_F + \mathfrak{n}_F$$

$$W_F = \{w \in W \mid wH = H \ \forall H \in \mathfrak{a}_F\} .$$

Note that every $\lambda \in \langle F\rangle$ has restriction 0 on $\mathfrak{a}_F$, and that the restrictions of $\langle F\rangle$ to $\mathfrak{a}(F)$ precisely form the roots of $\mathfrak{a}(F)$ in $\mathfrak{m}_F$. By Bourbaki [a] V §3.3, Proposition 2, W_F is generated by the reflections in $\langle F\rangle$, and is therefore identical to the Weyl group of the root system $\langle F\rangle$ on $\mathfrak{a}(F)$.

Let A_F, $A(F)$, A^F, N_F, $\overline{N}_F$, $N(F)$, $\overline{N}(F)$, $M_{F,0}$, $P_{F,0}$ and $B_{F,0}$ be the analytic subgroups of G with Lie algebras $\mathfrak{a}_F$, $\mathfrak{a}(F)$ etc., and let $M_F = M_{F,0} \cdot M$, $P_F = P_{F,0} \cdot M$, $B_F = B_{F,0} \cdot M$. (It is easily seen that M normalizes $\mathfrak{m}_F$, $\mathfrak{p}_F$ and $\mathfrak{b}_F$). For $w \in W$ let $\overline{w} = wW_F \in W/W_F$.

Proposition 3.2.2

(i) P_F is a standard parabolic subgroup for each $F \subset \Delta$, and $F \to P_F$ establishes a bijection between subsets of Δ and standard parabolic subgroups.

(ii) $P_F = M_F A_F N_F$ ("Langlands decomposition") and $M_F \times A_F \times N_F \ni (m,a,n) \longrightarrow man \in P_F$ is an analytic diffeomorphism.

(iii) M_F is a real reductive Lie group with compact center and it meets every connected component of G . Its Iwasawa decomposition is

$$M_F = (M_F \cap K)A(F)N(F) \quad .$$

(iv) $M_F A_F$ is the centralizer of $\mathfrak{a}_F$ in G , and normalizes N_F . P_F is its own normalizer in G .

(v) $G = KP_F$.

(vi) $\overline{N}_F P_F$ is open and dense in G and

$$\overline{N}_F \times P_F \ni (\overline{n}, p) \longrightarrow \overline{n}p \in \overline{N}_F P_F$$

is an analytic diffeomorphism.

(vii) $G = \bigcup_{\overline{w} \in W/W_F} N_F w P_F$ (disjoint union).

(viii) B_F is a closed subgroup of G .

(ix) If $E \subset F \subset \Delta$ then $P_E \subset P_F$, $M_E \subset M_F$, $A_E \supset A_F$ and $N_E \supset N_F$. $P_E \cap M_F$ is a parabolic subgroup of M_F with Langlands decomposition $P_E \cap M_F = M_E(A_E \cap A(F))(N_E \cap N(F))$.

Proof: See the references mentioned at the beginning of this section. □

Note that $A = A_F A^F = A_F A(F)$ and $\overline{N} = \overline{N}_F \overline{N}(F)$. Combining this with (ii), (iii) and (vi) of the proposition above, we get that

$$(3.13) \qquad \overline{N} \times A^F \times B_F \ni (\overline{n}, a, b) \longrightarrow \overline{n}ab \in \overline{N}_F P_F$$

is an analytic diffeomorphism onto an open dense submanifold of G .

3.3 Invariant differential operators

Let G be a real reductive Lie group with compact center and let D be a differential operator on the manifold $X = G/K$. D is called invariant if it commutes with the transformations $xK \to gxK$ for all $g \in G$.

We denote by $\mathbb{D}(G/K)$ the algebra of all invariant differential operators on G/K . A priori, there is some incorrectness in this notation, since it is not obvious that $\mathbb{D}(G/K)$ only depends on the space G/K. However, as we shall see below, every differential operator on $G/K \simeq G_1/K \cap G_1$ invariant for G_1 is invariant for G (the converse is obvious). Therefore the notation is correct.

Let $U(\mathfrak{g})$ be the enveloping algebra of $\mathfrak{g}_c$, and denote by $U(\mathfrak{g})^K$ the centralizer of K in $U(\mathfrak{g})$. Note that $U(\mathfrak{g})^K$ is also the centralizer of $\mathfrak{k}$ in $U(\mathfrak{g})$ by (3.10). There is a canonical homomorphism

$$\Gamma : U(\mathfrak{g})^K \longrightarrow \mathbb{D}(G/K)$$

coming from the right action of $U(\mathfrak{g})$ on G

$$(X_r f)(g) = \frac{d}{dt} f(g \exp tX)\Big|_{t=0}$$

for $f \in C^\infty(G)$, $g \in G$, $X \in \mathfrak{g}$.

By Helgason [n] Ch. 2, Theorem 4.6 every differential operator on $G_1/K \cap G_1$ invariant for G_1 is in the image of Γ, and hence it is invariant for G (as claimed above). Thus Γ is surjective onto $\mathbb{D}(G/K)$, and on the other hand its kernel is $U(\mathfrak{g})^K \cap U(\mathfrak{g})\mathfrak{k}_c$ (cf. loc. cit.), whence

$$\mathbb{D}(G/K) \approx U(\mathfrak{g})^K / U(\mathfrak{g})^K \cap U(\mathfrak{g})\mathfrak{k}_c .$$

From the Iwasawa decomposition $\mathfrak{g} = \overline{\mathfrak{n}} \oplus \mathfrak{a} \oplus \mathfrak{k}$ and the Poincaré-Birkhoff-Witt theorem it follows that

$$U(\mathfrak{g}) = \overline{\mathfrak{n}}_c U(\mathfrak{g}) \oplus U(\mathfrak{a}) \oplus U(\mathfrak{g})\mathfrak{k}_c . \tag{3.14}$$

Let δ be the projection of $U(\mathfrak{g})$ to $U(\mathfrak{a})$ with respect to this decomposition. Restricting δ to $U(\mathfrak{g})^K$ it can be seen that δ is a homomorphism and that its kernel in $U(\mathfrak{g})^K$ is precisely $U(\mathfrak{g})^K \cap U(\mathfrak{g})\mathfrak{k}_c$ (Loc. cit. Ch. 2, Theorem 5.17). To give a nice description of the image of $U(\mathfrak{g})^K$ by δ, we introduce the algebra automorphism η of $U(\mathfrak{a})$ generated by $\eta(X) = X - \rho(X)$, $X \in \mathfrak{a}$ and put $\gamma = \eta \circ \delta$. Then $\gamma(U(\mathfrak{g})^K)$ equals the set $U(\mathfrak{a})^W$ of Weyl group invariant elements of $U(\mathfrak{a})$ (cf. loc. cit.). Thus we also have an isomorphism:

$$\gamma : U(\mathfrak{g})^K / U(\mathfrak{g})^K \cap U(\mathfrak{g})\mathfrak{k}_c \xrightarrow{\sim} U(\mathfrak{a})^W .$$

Since $\mathfrak{a}$ is abelian we can identify $U(\mathfrak{a})$ with the polynomials in $\mathfrak{a}_c^*$. By Chevalley's theorem (Bourbaki [a]V §5.5, Theorem 3), $U(\mathfrak{a})^W$ is a polynomial ring generated by n algebraically independent homogeneous elements $p_1,\dots,p_n \in U(\mathfrak{a}) \simeq \mathbb{C}[H_1,\dots,H_n]$ and 1.

For each $\lambda \in \mathfrak{a}_c^*$ we define an algebra homomorphism χ_λ of $\mathbb{D}(G/K)$ to $\mathbb{C}$ by $\chi_\lambda(\Gamma(u)) = \delta(u)(\lambda - \rho) = \gamma(u)(\lambda)$ for each $u \in U(\mathfrak{g})^{\mathfrak{k}}$. It follows from the results described above that $\{\chi_\lambda \mid \lambda \in \mathfrak{a}_c^*\}$ constitutes all algebra homomorphisms from $\mathbb{D}(G/K)$ to $\mathbb{C}$, and that $\chi_\lambda = \chi_\mu$ if and only if there exists a $w \in W$ such that $\lambda = w\mu$.

Notice that if instead of (3.14) we use the decomposition

$$U(\mathfrak{g}) = \mathfrak{k}_c U(\mathfrak{g}) \oplus U(\mathfrak{a}) \oplus U(\mathfrak{g})\mathfrak{n}_c \tag{3.15}$$

to define the projection $U(\mathfrak{g}) \longrightarrow U(\mathfrak{a})$, then on $U(\mathfrak{g})^K$ this projection coincides with δ.

It is of importance, for each $F \subset \Delta$, to relate the algebra $\mathbb{D}(M_F/M_F \cap K)$ of invariant differential operators on $M_F/M_F \cap K$ to $\mathbb{D}(G/K)$. This is most conveniently done by considering the algebra $U(\mathfrak{m}_F + \mathfrak{a}_F)^{M_F \cap K}$. Let $\mathfrak{q}_F$ denote the orthocomplement of $\mathfrak{m}_F \cap \mathfrak{k}$ in $\mathfrak{k}$. Then $\mathfrak{g}$ decomposes as $\mathfrak{g} = \overline{\mathfrak{n}}_F \oplus \mathfrak{a}_F \oplus \mathfrak{m}_F \oplus \mathfrak{q}_F$ and it follows that

$$U(\mathfrak{g}) = (\overline{\mathfrak{n}}_F)_c U(\mathfrak{g}) \oplus U(\mathfrak{a}_F + \mathfrak{m}_F) \oplus U(\mathfrak{g})(\mathfrak{q}_F)_c \,. \tag{3.16}$$

Let δ_F be the projection of $U(\mathfrak{g})$ to $U(\mathfrak{m}_F + \mathfrak{a}_F)$ with respect to this decomposition. Since $\mathfrak{m}_F \cap \mathfrak{k}$ normalizes $\overline{\mathfrak{n}}_F$ and $\mathfrak{q}_F$, δ_F maps $U(\mathfrak{g})^K$ into $U(\mathfrak{m}_F + \mathfrak{a}_F)^{M_F \cap K}$. Moreover we see that the image of $U(\mathfrak{g})\mathfrak{k}_c$ by δ_F is contained in $U(\mathfrak{m}_F + \mathfrak{a}_F)(\mathfrak{m}_F \cap \mathfrak{k}_c)$, and hence δ_F induces a mapping

$$\mathbb{D}(G/K) \longrightarrow \mathbb{D}(M_F A_F/M_F \cap K) \,. \tag{3.17}$$

The map (3.17) is in general not surjective, as can be seen from the following considerations. Let δ^F be the projection of $U(\mathfrak{m}_F + \mathfrak{a}_F)$ to $U(\mathfrak{a})$ with respect to the decomposition

$$U(\mathfrak{m}_F + \mathfrak{a}_F) = (\overline{\mathfrak{n}}(F))_c U(\mathfrak{m}_F + \mathfrak{a}_F) \oplus U(\mathfrak{a}) \oplus U(\mathfrak{m}_F + \mathfrak{a}_F)(\mathfrak{m}_F \cap \mathfrak{k})_c$$

Then δ^F is the analog for $\mathfrak{m}_F + \mathfrak{a}_F$ of δ. Obviously $\delta = \delta^F \circ \delta_F$. Let η_F be the algebra automorphism of $U(\mathfrak{m}_F + \mathfrak{a}_F)$ generated by $\eta_F(X+H) = X + H - \rho(H)$ for $X \in \mathfrak{m}_F$, $H \in \mathfrak{a}_F$, and let $\gamma_F = \eta_F \circ \delta_F$. Also let $\eta^F : U(\mathfrak{a}) \longrightarrow U(\mathfrak{a})$ be the analog for $\mathfrak{m}_F + \mathfrak{a}_F$ of η for $\mathfrak{g}$ (i.e., $\eta^F(X) = X - \rho_F(X)$ where $\rho_F = \frac{1}{2} \sum_{\alpha \in \langle F \rangle \cap \Sigma^+} m(\alpha)\alpha$), and let $\gamma^F = \eta^F \circ \delta^F : U(\mathfrak{m}_F + \mathfrak{a}_F) \to U(\mathfrak{a})$.

It follows easily that

$$(3.18) \qquad \gamma = \gamma^F \circ \gamma_F$$

(note that $\rho(X) = \rho_F(X)$ for $X \in \mathfrak{a}(F)$ since the adjoint action of M_F on $\mathfrak{n}_F$ has determinant 1).

Now $\gamma(U(\mathfrak{g})^K) = U(\mathfrak{a})^W$ whereas $\gamma^F(U(\mathfrak{m}_F + \mathfrak{a}_F)^{M_F \cap K}) = U(\mathfrak{a})^{W_F}$, showing that γ_F - and hence (3.17) - is not surjective. From the theory of finite reflexion groups it follows that we can select elements $p_1, \ldots, p_r \in U(\mathfrak{a})^{W_F}$ such that

$$U(\mathfrak{a})^{W_F} = \Sigma_{i=1}^{r} p_i \, U(\mathfrak{a})^W .$$

Here $r = |W/W_F|$ and we can take $p_1 = 1$. Hence there are elements $\nu_1, \ldots, \nu_r \in U(\mathfrak{m}_F + \mathfrak{a}_F)^{M_F \cap K}$ with $\nu_1 = 1$ such that

$$(3.19) \quad U(\mathfrak{m}_F + \mathfrak{a}_F)^{M_F \cap K} \subset \Sigma_{i=1}^{r} \nu_i \delta_F(U(\mathfrak{g})^K) + U(\mathfrak{m}_F + \mathfrak{a}_F)(\mathfrak{m}_F \cap \mathfrak{h})_c$$

(cf. Harish-Chandra [c]I p. 250, Lemma 8, and II p. 564, Lemma 15).

3.4 Notes

This chapter only contains standard material for which references have been given already.

4. A compact imbedding

Let X be a Riemannian symmetric space. It is the purpose of this chapter to construct an imbedding of X into a compact real analytic manifold $\tilde{X}$. For the study of the asymptotic behavior of functions on X, which we shall carry out in the next chapters, this is of crucial importance.

The fundamental example is the upper half plane realization of $X = SL(2,\mathbb{R})/SO(2)$ which is constructed as follows: Let $\tilde{X} = \mathbb{C}\mathbb{P}^1 = \mathbb{C} \cup \{\infty\}$ be the one dimensional complex projective space. There is a natural action of $G = SL(2,\mathbb{R}) = \{\begin{pmatrix} a & b \\ c & d \end{pmatrix} \mid ad-bc = 1\}$ on this space, viz.

$$\begin{pmatrix} a & b \\ c & d \end{pmatrix} z = \frac{az + b}{cz + d} .$$

The orbits are $X_{\pm} = \{z = x + \sqrt{-1}\, y \mid \pm y > 0\}$ and $X_0 = \mathbb{R} \cup \{\infty\}$. The isotropy group of $\sqrt{-1}$ (and of $-\sqrt{-1}$) is $SO(2)$ and so $X_+ \simeq X_- \simeq X$. Thus we have an imbedding of X into the compact manifold $\tilde{X}$ (in fact we have two such imbeddings). Notice that the boundary of X in $\tilde{X}$ is $X_0 \simeq G/P$, where $P = \{$lower triangular matrices in $G\}$.

It is the generalization to arbitrary Riemannian symmetric spaces of this imbedding of the hyperbolic space $SL(2,\mathbb{R})/SO(2)$, which we present here. In short, it is a patchwork consisting of 2^n copies of X with the different homogeneous spaces G/B_F $(F \subsetneq \Delta)$ appearing on the various boundaries between the X's.

4.1 Construction and analytic structure of $\tilde{X}$

Let G be a real reductive Lie group with compact center, and let notation be as in Chapter 3. First some notation:

For $s \in \mathbb{R}$, define $\operatorname{sgn} s$ to be 1 if $s > 0$, 0 if $s = 0$

and -1 if $s < 0$. For $t = (t_1,\dots,t_n) \in \mathbb{R}^n$ let

$$\operatorname{sgn} t = (\operatorname{sgn} t_1,\dots,\operatorname{sgn} t_n) \in \{-1,0,1\}^n .$$

For $\varepsilon \in \{-1,0,1\}^n$, let

$$\mathbb{R}^n_\varepsilon = \{t \in \mathbb{R}^n \mid \operatorname{sgn} t = \varepsilon\}$$

then $\mathbb{R}^n = \bigcup_{\varepsilon \in \{-1,0,1\}^n} \mathbb{R}^n_\varepsilon$ (disjoint union).

For $t = (t_1,\dots,t_n) \in \mathbb{R}^n$ we define a subset F_t of the set Δ of simple roots for Σ^+ by

$$F_t = \{\alpha_j \in \Delta \mid t_j \neq 0\} .$$

From Section 3.2 we then have the subgroups P_{F_t}, B_{F_t} etc. of G, which we write P_t, B_t etc. for short. Thus, for example, $P_0 = P$ and $P_{(1,\dots,1)} = G$.

Define a map $\mathbb{R}^n \ni t \longrightarrow a_t \in A^{F_t}$ by $a_0 = e$ and

$$a_t = \exp(-\Sigma_{\alpha_j \in F_t} \log|t_j| H_j) \tag{4.1}$$

for $t \neq 0$. Via this map, $\mathbb{R}^n_\varepsilon$ is isomorphic to A^{F_ε} for each $\varepsilon \in \{-1,0,1\}^n$.

We are now ready for the definition of $\widetilde{X}$. We say that two elements (g,t) and (g',t') of $G \times \mathbb{R}^n$ are equivalent if $\operatorname{sgn} t = \operatorname{sgn} t'$ and moreover

$$ga_t \in g'a_{t'}B_t .$$

Since $B_t = B_{t'}$ when $\operatorname{sgn} t = \operatorname{sgn} t'$, this is in fact an equivalence relation, which we write $(g,t) \sim (g',t')$. Let the topological space $\widetilde{X}$ be the quotient space $G \times \mathbb{R}^n/\sim$ for this relation.

Let $\pi : G \times \mathbb{R}^n \longrightarrow \widetilde{X}$ be the projection. For $\varepsilon \in \{1,0,-1\}^n$, let

$$\mathcal{O}_\varepsilon = \pi(G \times \mathbb{R}^n_\varepsilon) .$$

The space $\widetilde{X}$ inherits from $G \times \mathbb{R}^n$ a continuous action of G, given by $x\pi(g,t) = \pi(xg,t)$. Then it is easy to see that

$$\widetilde{X} = \bigcup_{\varepsilon \in \{-1,0,1\}^n} \mathcal{O}_\varepsilon \quad \text{(disjoint union)} \tag{4.2}$$

gives the orbital decomposition of $\widetilde{X}$ for this action.

For each $\varepsilon \in \{-1,0,1\}^n$ we define a map $G/B_\varepsilon \longrightarrow O_\varepsilon$ by $gB_\varepsilon \longrightarrow \pi(g,\varepsilon)$. Since $a_\varepsilon = e$, this map is well defined and in fact establishes a bijective homeomorphism which is equivariant for the actions of G :

$$G/B_\varepsilon \xrightarrow{\sim} O_\varepsilon \quad .$$

In particular, G/K can be identified with each of the 2^n open orbits O_ε where $\varepsilon \in \{-1,1\}^n$, and the union of these open orbits is dense in $\widetilde{X}$. From the Cartan decomposition (3.3) it follows that $O_\varepsilon \subset \pi(K \times [-1,1]^n)$ for each $\varepsilon \in \{-1,1\}^n$, whence $\widetilde{X}$ equals the (quasi-)compact set $\pi(K \times [-1,1]^n)$. (We have not yet seen that $\widetilde{X}$ is Hausdorff. This follows later.)

$\widetilde{X}$ is connected since B_F meets every connected component of G for all $F \subset \Delta$. In fact it is easily seen that $\widetilde{X}$ is independent of the choice of group G with $X = G/K$. We have thus proved:

<u>Proposition 4.1.1</u> $\widetilde{X}$ <u>is a compact connected</u> G-<u>space with the orbital decomposition</u> (4.2). <u>Each</u> O_ε <u>is homeomorphic to</u> G/B_ε . <u>In particular,</u> $O_0 \simeq G/P$, <u>and</u> $O_\varepsilon \simeq X$ <u>when</u> $\varepsilon \in \{-1,1\}^n$, <u>and</u> $\bigcup_{\varepsilon \in \{-1,1\}^n} O_\varepsilon$ <u>is dense in</u> $\widetilde{X}$.

We shall define on $\widetilde{X}$ a structure of a real analytic manifold such that the action of G is analytic. Before we do this we need a lemma, which will serve to ensure analyticity of changes of coordinates, and a second lemma necessary for the proof of the first lemma.

Let $\Omega_g \subset \overline{N} \times \mathbb{R}^n$ for $g \in G$ be the set

$$\Omega_g = \{(\overline{n},t) \mid g\overline{n} \in \overline{N}_t P_t\} \quad .$$

Notice that if $t_j \neq 0$ for all j , then $(\overline{n},t) \in \Omega_g$ for all $\overline{n} \in \overline{N}$, hence Ω_g is dense in $\overline{N} \times \mathbb{R}^n$. Obviously, $\Omega_e = \overline{N} \times \mathbb{R}^n$.

For $g \in G$ let the map

$$\psi_g : \Omega_g \longrightarrow \overline{N} \times \mathbb{R}^n$$

be given by $\psi_g(\overline{n},t) = (\overline{n}',t')$ where $(\overline{n}',t')$ by (3.13) is determined uniquely by $(\overline{n}',t') \sim (g\overline{n},t)$.

From the transitivity of $\sim$ it follows immediately that

$$\psi_{g'} \circ \psi_g = \psi_{g'g} \tag{4.3}$$

on $\Omega_{g'g} \cap \Omega_g$ for $g', g \in G$.

<u>Lemma 4.1.2 Let</u> $g \in G$. <u>Then</u> Ω_g <u>is open in</u> $\bar{N} \times \mathbb{R}^n$ <u>and</u> ψ_g <u>is an analytic diffeomorphism of</u> Ω_g <u>onto</u> $\Omega_{g^{-1}}$.

<u>Proof</u>: From (4.3) it follows easily that ψ_g is a bijection of Ω_g onto $\Omega_{g^{-1}}$. For the proof of the remaining part of the lemma, it is convenient to introduce a map Ψ which unites all ψ_g $(g \in G)$.

Let $\Omega \subset G \times \mathbb{R}^n$ be the dense set

$$\Omega = \cup_{g \in G} g\Omega_g = \{(g,t) \mid g \in \bar{N}_t P_t\}$$

then Ω is the disjoint union

$$\bigcup_{\varepsilon \in \{-1,0,1\}^n} \bar{N}_\varepsilon P_\varepsilon \times \mathbb{R}^n_\varepsilon \quad .$$

We define the map

$$\Psi : \Omega \longrightarrow \bar{N} \times \mathbb{R}^n$$

by $\Psi(g,t) = \psi_g(e,t)$, i.e., $\Psi(g,t) = (\bar{n}',t')$ where $(\bar{n}',t')$ is determined uniquely by $(\bar{n}',t') \sim (g,t)$.

We claim that Ω is open and that Ψ is real analytic on Ω. This claim implies the statement of the lemma, since $\Omega_g = g^{-1}\Omega \cap (\bar{N} \times \mathbb{R}^n)$ and $\psi_g(\bar{n},t) = \Psi(g\bar{n},t)$.

Obviously, each $\bar{N}_\varepsilon P_\varepsilon \times \mathbb{R}^n_\varepsilon$ is an open subset of $G \times \mathbb{R}^n_\varepsilon$, and the restriction of Ψ to there is analytic. The only problem is whether these pieces correspond together nicely.

Let $(g_o, t_o) \in \Omega$ be given. Assume first $g_o = \bar{n}_o \in \bar{N}$. Fix $Y \in \mathfrak{g}$ and consider for each $\varepsilon \in \{-1,0,1\}^n$ the vector field Y_ε on $\bar{N} \times \mathbb{R}^n_\varepsilon$ corresponding to the action of $\exp Y$ on G/B_ε via the injection

$$\bar{N} \times \mathbb{R}^n_\varepsilon \ni (\bar{n},t) \longrightarrow \bar{n}a_t B_\varepsilon \in G/B_\varepsilon \, . \tag{4.4}$$

By Lemma 4.1.3 below, we see that the vector fields Y_ε piece analytically together, and thus determine an analytic vector field on $\bar{N} \times \mathbb{R}^n$.

By Lie's theorem (Varadarajan [a], Theorem 2.16.8) let φ_g be the corresponding local transformation group of $\bar{N} \times \mathbb{R}^n$. Then we see that for each point $p \in \bar{N} \times \mathbb{R}^n$ there is a neighborhood U_p of e in G such that $p \in \Omega_g$ and $\psi_g(p) = \varphi_g(p)$ for $g \in U_p$.

Let $U \subset G$, $V \subset \bar{N}$ and $W \subset \mathbb{R}^n$ be open neighborhoods of e, $\bar{n}_o$ and t_o, respectively, such that φ_g is defined in $V \times W$ for all $g \in U$. Since $\bar{n}_o \in \bar{N}P$ and $\bar{N}P$ is open we may assume that $UV \subset \bar{N}P$. This implies that $(\bar{n}, t) \in \Omega_g$ for all $g \in U$, $\bar{n} \in V$ and $t \in \mathbb{R}^n$.

Assume that U is connected. We claim that then $\psi_g = \varphi_g$ on $V \times W$ for all $g \in U$. To prove this, pick $(\bar{n}, t) \in V \times W$ and consider the set

$$\{g \in U \mid \psi_g(\bar{n}, t) = \varphi_g(\bar{n}, t)\} .$$

This set is open, because if it contains g then it contains $U_p g$ where $p = \psi_g(\bar{n}, t)$. This follows from (4.3). On the other hand, it is closed since $g \longrightarrow \psi_g(\bar{n}, t)$ is continuous. Since the set contains e, it equals U, as claimed.

Since $(\bar{n}_o, t) \in \Omega_g$ for $g \in U$ and $t \in W$, we have $(g\bar{n}_o, t) \in \Omega$ for $(g, t) \in U \times W$. Furthermore $\Psi(g\bar{n}_o, t) = \Psi_g(\bar{n}_o, t) = \varphi_g(\bar{n}_o, t)$ is analytic for $(g, t) \in U \times W$. Thus our claim on Ω and Ψ holds in a neighborhood of $(\bar{n}_o, t_o)$ in $G \times \mathbb{R}^n$.

Assume next that $g_o \in \bar{N}_F P_{F,0}$ where $F = F_{t_o}$. Since $\bar{N}_F P_{F,0}$ is connected there is a continuous curve $\zeta : [0,1] \longrightarrow G$ inside this open set with $\zeta(0) = e$ and $\zeta(1) = g_o$. By what we have proved, there is a neighborhood U in $G \times \mathbb{R}^n$ of the compact set $\{\Psi(\zeta(s), t_o) \mid s \in [0,1]\}$, such that Ψ is defined and analytic in U.

Now, by (4.3), we have that $\Psi(g'\Psi(g,t)) = \Psi(g'g, t)$ and hence

$$\Psi(g,t) = \Psi(\zeta(1)\zeta(1-\tfrac{1}{j})^{-1}\Psi(\zeta(1-\tfrac{1}{j})\zeta(1-\tfrac{2}{j})^{-1}\ldots\Psi(\zeta(\tfrac{1}{j})g_o^{-1}g, t)\ldots))$$

for $j \in \mathbb{N}$. Taking a sufficiently large j, this proves that Ψ is defined and analytic in a neighborhood of (g_o, t_o).

Finally, if $g_o \in \bar{N}_F P_F$, then $g_o = mg$, where $m \in M$ and $g \in \bar{N}_F P_{F,0}$. Since $\Psi(m\bar{n}, t) = (m\bar{n}m^{-1}, t)$ for all $(\bar{n}, t) \in \bar{N} \times \mathbb{R}^n$ and since Ψ is defined and analytic in a neighborhood of (g, t_o), it follows from the identity $\Psi(mg, t) = \Psi(m\Psi(g, t))$ that Ψ is defined and analytic in a neighborhood of (g_o, t_o).

Except for the proof of the next lemma, the proof of Lemma 4.1.2 is complete. □

<u>Lemma 4.1.3</u> <u>For</u> $Y \in \mathfrak{g}$ <u>and</u> $\varepsilon \in \{-1,0,1\}^n$ <u>the vector field</u> Y_ε <u>on</u> $\bar{N} \times \mathbb{R}^n_\varepsilon$ <u>corresponding to the action of</u> $\exp Y$ <u>on</u> G/B_ε <u>via</u> (4.4) <u>is determined by</u>

$$Y_\varepsilon = \left[\Sigma_{\alpha \in \Sigma^{+\prime}} (c_\alpha(\bar{n}) t^{2\alpha} + c_{-\alpha}(\bar{n})) X_{-\alpha} - \Sigma_{i=1}^n c_i(\bar{n}) t_i \frac{\partial}{\partial t_i} \right] \Bigg|_{\bar{N} \times \mathbb{R}^n_\varepsilon}$$

<u>where</u> $t^{2\alpha} = \prod_{j=1}^n t_j^{2\alpha(H_j)}$ <u>and where</u> $c_{\pm\alpha}$ <u>and</u> c_i <u>are analytic functions on</u> $\bar{N}$ <u>determined by</u>

$$\text{(4.5)} \qquad \operatorname{Ad} \bar{n}^{-1} Y \equiv \Sigma_{\alpha \in \Sigma^{+\prime}} (c_\alpha(\bar{n}) X_\alpha + c_{-\alpha}(\bar{n}) X_{-\alpha}) + \Sigma_{i=1}^n c_i(\bar{n}) H_i$$

<u>modulo</u> $\mathfrak{m}$.

<u>Proof:</u> According to (3.13) we write for s sufficiently small

$$\exp(sY)\bar{n}a_t \in \bar{n} \exp \bar{N}(s) a_t A(s) B_t$$

where $\bar{N}(s) \in \bar{\mathfrak{n}}$ and $A(s) \in \mathfrak{a}^{F_t}$.

Multiplying from the left with $a_t^{-1} \bar{n}^{-1}$ and differentiating the expression with respect to s at $s = 0$ gives

$$\operatorname{Ad} a_t^{-1} \operatorname{Ad} \bar{n}^{-1} Y \equiv \operatorname{Ad} a_t^{-1} \frac{d\bar{N}}{ds}(0) + \frac{dA}{ds}(0) \bmod \mathfrak{b}_t \quad .$$

It is easy to verify that for $t \in \mathbb{R}^n_\varepsilon$ and $\alpha \in \Sigma^{+\prime}$:

$$\operatorname{Ad} a_t^{-1} X_\alpha \equiv t^{2\alpha} \operatorname{Ad} a_t^{-1} X_{-\alpha} \bmod \mathfrak{b}_t \quad .$$

(Note that if $X_\alpha \in \mathfrak{n}_{F_t}$, then $t^{2\alpha} = 0$, since there is an $i \notin F_t$ with $\alpha(H_i) \neq 0$).

Inserting this into the equation obtained from (4.5) by applying $\operatorname{Ad} a_t^{-1}$ gives formulas for $\frac{d\bar{N}}{ds}(0)$ and $\frac{dA}{ds}(0)$ as follows:

$$\frac{d\bar{N}}{ds}(0) = \Sigma_{\alpha \in \Sigma^{+\prime}} (c_\alpha(\bar{n}) t^{2\alpha} + c_{-\alpha}(\bar{n})) X_{-\alpha}$$

and

$$\frac{dA}{ds}(0) = \Sigma_{i=1}^n c_i(\bar{n}) H_i \, .$$

Since H_i acting on A^{F_t} corresponds to the operator $-t_i \frac{\partial}{\partial t_i}$ under (4.1), the lemma follows. □

4.1.4 Example Let $G = SL(2,\mathbb{R})$, $K = SO(2)$,

$$A = \left\{ \begin{pmatrix} a & 0 \\ 0 & a^{-1} \end{pmatrix} \;\middle|\; a > 0 \right\} \quad \text{and} \quad \overline{N} = \left\{ \overline{n}_x = \begin{pmatrix} 1 & x \\ 0 & 1 \end{pmatrix} \;\middle|\; x \in \mathbb{R} \right\} .$$

In this example $a_t = \begin{pmatrix} |t|^{1/2} & 0 \\ 0 & |t|^{-1/2} \end{pmatrix}$ for $t \neq 0$. With $g = \begin{pmatrix} a & b \\ c & d \end{pmatrix}$ we easily get that

$$\Omega_g = \{(\overline{n}_x, t) \mid (x,t) \neq (-\tfrac{d}{c}, 0)\}.$$

Moreover, we can determine ψ_g explicitly : if $t \neq 0$ we have to use the Iwasawa decomposition $G = \overline{N}AK$ on $g\overline{n}_x$, and if $t = 0$ we have to use Bruhat decomposition $\overline{N}MAN$ on this element with $x \neq -\frac{d}{c}$. The result is

$$\psi_g(x,t) = \begin{cases} \left(\dfrac{act^2 + (ax+b)(cx+d)}{c^2t^2 + (cx+d)^2} , \dfrac{t}{c^2t^2 + (cx+d)^2} \right) & \text{if } t \neq 0 \\[2ex] \left(\dfrac{ax+b}{cx+d} , 0 \right) & \text{if } t = 0 , \ cx + d \neq 0 \end{cases}$$

(identifying $\overline{n}_x$ with x).

Thus ψ_g is analytic on Ω_g in accordance with Lemma 4.1.2. □

We are now ready to define the analytic structure on $\widetilde{X}$.

For each $g \in G$, define $\varphi_g : \overline{N} \times \mathbb{R}^n \longrightarrow \widetilde{X}$ by

$$(4.6) \qquad \varphi_g(\overline{n}, t) = \pi(g\overline{n}, t) .$$

From (3.13) and the definition of $\widetilde{X}$ it follows that φ_g is injective and continuous. For an open subset $V \subset \overline{N} \times \mathbb{R}^n$, $\varphi_g(V)$ is open in $\widetilde{X}$ since its preimage in $G \times \mathbb{R}^n$ is precisely $g\Psi^{-1}(V)$ where Ψ is given in the proof of Lemma 4.1.2. Hence φ_g is a homeomorphism onto an open subset of $\widetilde{X}$. Moreover we have

$$\widetilde{X} = \cup_{g \in G} \varphi_g(\overline{N} \times \mathbb{R}^n) .$$

For $g, g' \in G$ we have

$$\varphi_g^{-1}(\varphi_{g'}(\overline{N} \times \mathbb{R}^n)) = \Omega_{g'^{-1}g}$$

and

$$\varphi_{g'}^{-1} \circ \varphi_g = \psi_{g'^{-1}g} \quad \text{on} \quad \Omega_{g'^{-1}g} \quad .$$

Therefore, by Lemma 4.1.2, $(\varphi_g)_{g \in G}$ is an atlas of a structure on $\tilde{X}$ of a real analytic manifold (we identify $\bar{N}$ with $\mathbb{R}^k$, where $k = \dim \bar{N}$, via the exponential map).

Notice that any two points of $\tilde{X}$ can be covered by a single coordinate patch, since for each $x \in \tilde{X}$ the set $\{g \in G \mid x \in \varphi_g(\bar{N} \times \mathbb{R}^n)\}$ is open and dense in G. Therefore, $\tilde{X}$ is Hausdorff.

It is obvious from (4.6) that the action of G on $\tilde{X}$ is real analytic. Therefore the orbits O_ε ($\varepsilon \in \{-1,0,1\}^n$) are analytic submanifolds. In fact if $x = \pi(g,\varepsilon) \in O_\varepsilon$ then O_ε in a neighborhood of x equals $\{\varphi_g(\bar{n},t) \mid \operatorname{sgn} t = \varepsilon\}$. Since

$$\bar{N} \times \mathbb{R}^n_\varepsilon \ni (\bar{n},t) \longrightarrow g\bar{n}a_t B_\varepsilon \in G/B_\varepsilon$$

is an analytic diffeomorphism onto an open subset of G/B_ε we also get that

$$O_\varepsilon \xrightarrow{\sim} G/B_\varepsilon$$

is an analytic diffeomorphism.

We have proved:

Theorem 4.1.5 *$\tilde{X}$ is a compact connected real analytic manifold on which G acts analytically. The orbits are diffeomorphic to the homogeneous spaces G/B_F, and G/B_F occurs precisely $2^{|F|}$ times. $\tilde{X}$ contains in particular 2^n copies of X as open subsets and one copy of G/P as a closed subset.*

The orbits O_ε ($\varepsilon \in \{0,1\}^n$, $\varepsilon \neq (1,\ldots,1)$) are called the *boundary orbits* of G in $\tilde{X}$. The boundary orbit $O_0 \cong G/P$ is called the *distinguished boundary* (orbit) in $\tilde{X}$.

For later purpose, we study ψ_g a little closer. For $g \in G$ we define a map $H_g : \Omega_g \longrightarrow \mathfrak{a}$ by

$$(4.7) \qquad g\bar{n}a_t \in \bar{N} \exp H_g(\bar{n},t)(M_t \cap K)N_t \quad .$$

Writing $\psi_g(\bar{n},t) = (\bar{n}^g, t^g)$ for $(\bar{n},t) \in \Omega_g$ we then get

$$t_i^g = \begin{cases} \operatorname{sgn} t_i \exp\langle -\alpha_i , H_g(\bar{n},t)\rangle & \text{if } t_i \neq 0 \\ 0 & \text{if } t_i = 0 \end{cases} . \tag{4.8}$$

<u>Lemma 4.1.6</u> <u>Let</u> $(\bar{n},t) \in \Omega_g$ <u>and assume that</u> $t_i = 0$. <u>Then</u>

$$\frac{\partial t_i^g}{\partial t_j}(\bar{n},t) = \delta_{ij} \exp\langle -\alpha_i , H_g(\bar{n},t)\rangle \quad . \tag{4.9}$$

<u>Proof</u>: The following identity follows from the definition of H_g :

$$H_{g'g}(\bar{n},t) = H_{g'}(\psi_g(\bar{n},t)) + H_g(\bar{n},t) - \log a_{t^g} \tag{4.10}$$

for $(\bar{n},t) \in \Omega_{g'g} \cap \Omega_g$.

For $g = e$, and more generally $g \in M$, the lemma is true since then $t_i^g = t_i$ and $H_g(\bar{n},t) = \log a_t$. By (4.10) and the chain rule, Lemma 4.1.6 holds for $g'g$ on $\Omega_{g'g} \cap \Omega_g$ if it holds for g' and g . Applying the final arguments of the proof of Lemma 4.1.2, we may therefore assume that $g = \exp Y$, $Y \in \mathfrak{g}$. Put $g_s = \exp sY$, $s \in \mathbb{R}$.

Let the map $\mathfrak{K} : \mathfrak{g} \longrightarrow \mathfrak{a}$ be the projection along $\bar{\mathfrak{n}} + \mathfrak{h}$, and note that $\bar{\mathfrak{n}} + \mathfrak{h} = \bar{\mathfrak{n}} + \mathfrak{m}_F \cap \mathfrak{h} + \mathfrak{n}_F$ for all $F \subset \Delta$. We claim that for arbitrary $t \in \mathbb{R}^n$:

$$\frac{d}{ds} H_{g_s}(\bar{n},t)\Big|_{s=0} = \mathfrak{K}(\operatorname{Ad} \bar{n}^{-1} Y) \quad . \tag{4.11}$$

To prove the claim, multiply (4.7) from the left with $(\bar{n}a_t)^{-1}$ and differentiate it with respect to s . This gives

$$\frac{d}{ds} H_{g_s}(\bar{n},t)\Big|_{s=0} = \mathfrak{K}(\operatorname{Ad} a_t^{-1} \operatorname{Ad} \bar{n}^{-1} Y)$$

but $\mathfrak{K}(\operatorname{Ad}(a)Z) = \mathfrak{K}(Z)$ for all $a \in A$, $Z \in \mathfrak{g}$, whence the claim.

Put $\psi_{g_s}(\bar{n},t) = (\bar{n}^s, t^s)$, then we get from (4.10) and (4.11) with $y = \exp\langle -\alpha_i , H_{g_s}(\bar{n},t)\rangle$:

$$\frac{dy}{ds} = \langle -\alpha_i , \mathfrak{K}(\operatorname{Ad}(\bar{n}^s)^{-1} Y)\rangle \, y \quad . \tag{4.12}$$

For $t_i > 0$, combining with (4.8), this gives that

$$\frac{d}{ds}(t_i^s) = \langle -\alpha_i, \mathfrak{h}(\mathrm{Ad}(\bar{n}^s)^{-1}Y)\rangle t_i^s$$

and then

$$\frac{d}{ds}\left(\frac{\partial t_i^s}{\partial t_j}\right) = \langle -\alpha_i, \mathfrak{h}(\mathrm{Ad}(\bar{n}^s)^{-1}Y)\rangle \frac{\partial t_i^s}{\partial t_j} - \langle \alpha_i, \frac{\partial}{\partial t_j}\mathfrak{h}(\mathrm{Ad}(\bar{n}^s)^{-1}Y)\rangle t_i^s .$$

Using Lemma 4.1.2, this implies by analyticity

$$\frac{d}{ds}\left(\frac{\partial t_i^s}{\partial t_j}\right) = \langle -\alpha_i, \mathfrak{h}(\mathrm{Ad}(\bar{n}^s)^{-1}Y)\rangle \frac{\partial t_i^s}{\partial t_j} \quad \text{at} \quad t_i = 0 .$$

Thus both sides of equation (4.9) for g_s satisfy the differential equation (4.12). We have already seen that they agree at $s = 0$. This proves the lemma. □

In the example 4.1.4, equation (4.9) is easily checked by computation.

4.2 Invariant differential operators on $\tilde{X}$

We will extend the action of $\mathbb{D}(G/K)$ from X to $\tilde{X}$. Let $\varepsilon \in \{-1,1\}^n$ and identify O_ε with G/K. Then each $D \in \mathbb{D}(G/K)$ defines a G-invariant differential operator D_ε on O_ε. Let τ_ε be the algebra automorphism of $U(\mathfrak{g})$ satisfying

$$\tau_\varepsilon(X) = \prod_{i=1}^{n} \varepsilon_i^{\alpha(H_i)} X$$

for any $\alpha \in \Sigma \cup \{0\}$ and $X \in \mathfrak{g}^\alpha$. Since τ_ε preserves $\mathfrak{h}$ and $\mathfrak{n}$, and is trivial on $\mathfrak{a}$, we have $\Gamma(\tau_\varepsilon u) = \Gamma(u)$ for all $u \in U(\mathfrak{g})^K$.

Proposition 4.2.1 For each $D \in \mathbb{D}(G/K)$ there is a unique G-invariant differential operator $\tilde{D}$ with analytic coefficients on $\tilde{X}$, whose restriction to O_ε for each $\varepsilon \in \{-1,1\}^n$ is D_ε. All G-invariant differential operators with analytic coefficients on $\tilde{X}$ are obtained in this way.

Proof: Let $D = \Gamma(u)$, $u \in U(\mathfrak{g})^K$. We find the expression of D_ε in the coordinates on $O_\varepsilon = \varphi_g(\bar{N} \times \mathbb{R}^n_\varepsilon)$ given by φ_g ($g \in G$ arbitrary). Let ι_g denote the isomorphism of $\bar{N}A$ with X by

$\iota_g(\bar{n}a) = \bar{n}a\,K$.

Let $u' \in U(\bar{\mathfrak{n}} + \mathfrak{a})$ be determined by $u' \equiv u \mod U(\mathfrak{g})\mathfrak{k}_c$. Then by the definition of Γ, D acts on $f \in C^\infty(X)$ as u' acts on $f \circ \iota_g \in C^\infty(\bar{N}A)$ from the right.

Let $Y = \sum_{\alpha \in \Sigma^+} c_{-\alpha} X_{-\alpha} + \sum_{i=1}^n c_i H_i \in \bar{\mathfrak{n}} + \mathfrak{a}$. Since $\mathrm{Ad}\, a\, X_{-\alpha} = a^{-\alpha} X_{-\alpha}$ for $a \in A$, the action of Y on the Lie group $\bar{N}A$ from the right is expressed as the following vector-field on $\bar{N} \times A$:

$$\sum_{\alpha \in \Sigma^+} c_{-\alpha}\, a^{-\alpha} X_{-\alpha} + \sum_{i=1}^n c_i H_i \quad . \tag{4.13}$$

By the isomorphism $\mathbb{R}^n_\varepsilon \ni t \longrightarrow a_t \in A$ ($\varepsilon \in \{-1,1\}^n$) it is easily seen that H_i corresponds to $-t_i \frac{\partial}{\partial t_i}$. Therefore we get the following expression of (4.13):

$$\sum_{\alpha \in \Sigma^+} c_{-\alpha} |t|^\alpha X_{-\alpha} - \sum_{i=1}^n c_i\, t_i \frac{\partial}{\partial t_i} \tag{4.14}$$

where $|t|^\alpha = \prod_{i=1}^n |t_i|^{\alpha(H_i)}$.

Applying τ_ε to Y we get instead of (4.14):

$$\sum_{\alpha \in \Sigma^+} c_{-\alpha}\, t^\alpha X_{-\alpha} - \sum_{i=1}^n c_i\, t_i \frac{\partial}{\partial t_i} \tag{4.15}$$

where $t^\alpha = \prod_{i=1}^n t_i^{\alpha(H_i)}$. But this expression has analytic coefficients in all of $\bar{N} \times \mathbb{R}^n$ (since $\alpha(H_i)$ is an integer), and is moreover independent of ε .

Since this holds for any $Y \in \bar{\mathfrak{n}} + \mathfrak{a}$, the similar statement holds for u' , and therefore the operators D_ε on O_ε extend to an analytic differential operator $\tilde{D}$ on $\tilde{X}$. $\tilde{D}$ is obviously unique, and since it is an analytic extension of an invariant differential operator it is itself invariant for G . The last statement is likewise obvious. □

4.3 Regular singularities

Fix $\lambda \in \mathfrak{a}_c^*$.

Theorem 4.3.1 *The system of differential equations on* $\tilde{X}$:

$$\mathcal{M}_\lambda : (\tilde{D} - \chi_\lambda(D))u = 0 \qquad \forall D \in \mathbb{D}(G/K)$$

has regular singularities along the walls $N_j = \pi(\{(g,t) \mid t_j = 0\})$ $(j = 1,\dots,n)$ *with the edge* $O_0 \simeq G/P$. *The characteristic exponents are given by* $s_w = (s_{s,1},\dots,s_{w,n}) \in \mathbb{C}^n$ *where* $w \in W$ *and*

$$s_{w,i} = \langle \rho - w\lambda , H_i \rangle \quad . \tag{4.16}$$

Proof: That $\tilde{D}$ has the form required by (I) in the definition (Section 2.2) follows immediately from (4.15). In fact, if $D = \Gamma(u)$ with $u \in U(\mathfrak{g})^K$, and

$$u = \Sigma_{p \in \mathbb{Z}_+^n,\, q \in \mathbb{Z}_+^k} a_{p,q} \cdot X_{-\beta_1}^{q_1} \dots\dots X_{-\beta_k}^{q_k} H_1^{p_1} \dots H_n^{p_n} \mod U(\mathfrak{g})\mathfrak{k}$$

where $\beta_1,\dots,\beta_k$ are elements of Σ^+ , then $\tilde{D}$ has the following coordinate expression

$$\tilde{D} = \Sigma_{p,q}\, a_{p,q} \left(t^{\beta_1} \frac{\partial}{\partial x_1}\right)^{q_1} \dots \left(t^{\beta_k} \frac{\partial}{\partial x_k}\right)^{q_k} \left(-t_1 \frac{\partial}{\partial t_1}\right)^{p_1} \dots \left(-t_n \frac{\partial}{\partial t_n}\right)^{p_n} \tag{4.17}$$

in the coordinates $(x,t) \longrightarrow \varphi_g(\exp(x_1 X_{-\beta_1} + \dots + x_k X_{-\beta_k}),t)$ for any $g \in G$.

It follows from (4.17) that the indicial polynomial of $\tilde{D}$ is

$$\Sigma_p\, a_{p,0}(-s_1)^{p_1} \dots (-s_n)^{p_n}$$

which is the same as $\delta(u)(-s)$. Therefore the indicial polynomial of $\tilde{D} - \chi_\lambda(D)$ equals

$$a(s) = \gamma(D)(\rho - s) - \gamma(D)(\lambda) \quad . \tag{4.18}$$

Now, the equation $p(\rho - s) = p(\lambda)$, $\forall p \in U(\mathfrak{a})^W$, implies $\rho - s = w\lambda$ for some $w \in W$, proving (4.16). Let $p_1, \ldots, p_n$ be the homogeneous generators for $U(\mathfrak{a})^W$, then

$$p_1(-s) = \ldots = p_n(-s) = 0$$

implies $s = 0$ since $p_1, \ldots, p_n$ are independent. Therefore property (II) holds for $\mathcal{M}_\lambda$. □

Notice that by the remark on Bezout's theorem in Section 2.2 it follows that

$$|W| = \prod_{i=1}^{n} \deg p_i$$

(cf. Bourbaki [a] V §5 n° 5.3 Corr.).

Since we want to take boundary values, we need Condition (B) of Section 2.5. That is provided for by the following proposition. Fix a coordinate system φ_g $(g \in G)$, and an i $(1 \leq i \leq n)$. Let $W(H_i)$ be the stabilizer of H_i in W and choose a set of representatives $w_1 = e, w_2, \ldots, w_m \in W$ for the right cosets of $W(H_i)$ in W (i.e., $m = |W|/|W(H_i)|$ and $W(H_i)\backslash W = \{w(H_i)w_1, \ldots, w(H_i)w_m\}$).

<u>Proposition 4.3.2</u> <u>There exist</u> m <u>elements</u> $D_1, \ldots, D_m$ <u>(depending on</u> i) <u>in</u> $\mathbb{D}(G/K)$ <u>such that the differential operator</u>

$$(4.19) \qquad Q_i = \Sigma_{j=1}^{m} \left(t_i \frac{\partial}{\partial t_i}\right)^{m-j} (\widetilde{D}_j - \chi_\lambda(D_j))$$

<u>has degree</u> m <u>and has regular singularities in the weak sense along</u> $\varphi_g(\bar{N} \times \{t_i = 0\})$. <u>Moreover the characteristic exponents</u> $s_j \in \mathbb{C}$ $(j = 1, \ldots, m)$ <u>of</u> Q_i <u>are given by</u> $s_j = \langle \rho - w_j\lambda, H_i \rangle$.

<u>Proof:</u> Let $F = \Delta \backslash \{\alpha_i\}$, then $W(H_i) = W_F$. Put

$$p = \prod_{j=1}^{m} (x + \rho(H_i) - w_j^{-1} H_i) \in U(\mathfrak{a})[x],$$

then it is easily seen that $p \in U(\mathfrak{a})^W[x]$, and therefore $p = \Sigma_{j=0}^{m} x^{m-j} p_j$ for some $p_j \in U(\mathfrak{a})^W$ of order $\leq j$ $(j = 0, \ldots, m)$.

Choose elements $u_j \in U(\mathfrak{g})^K$ of order $\leq j$ such that $\gamma(u_j) = p_j$. In particular we take $u_0 = 1$.

We claim that

$$(4.20) \qquad H_i^m + \Sigma_{j=1}^m H_i^{m-j} u_j \in \overline{\mathfrak{n}}_{F,c} U(\mathfrak{g}) + U(\mathfrak{g})\mathfrak{z}_c .$$

To prove this let

$$q = \prod_{j=2}^{m} (x + \rho(H_i) - w_j^{-1} H_i) ,$$

then $q \in U(\mathfrak{a})^{W_F}[x]$, and therefore $q = \Sigma_{j=0}^{m-1} x^{m-j-1} q_j$ for some $q_j \in U(\mathfrak{a})^{W_F}$ $(j = 0, \ldots, m-1)$. Choose elements $v_j \in U(\mathfrak{m}_F + \mathfrak{a}_F)^{K \cap M_F}$ such that $\gamma^F(v_j) = q_j$. Put

$$u(x) = \Sigma_{j=0}^m x^{m-j} u_j \in U(\mathfrak{g})^K[x]$$

and

$$v(x) = \Sigma_{j=0}^{m-1} x^{m-j-1} v_j \in U(\mathfrak{m}_F + \mathfrak{a}_F)^{K \cap M_F}[x] .$$

Then

$$\gamma^F(\gamma_F(u(x))) = \gamma(u(x)) = p = (x + \rho(H_i) - H_i) q = \gamma^F((x - H_i) v(x))$$

whence

$$\gamma_F(u(x)) \equiv (x - H_i) v(x) \bmod (U(\mathfrak{m}_F + \mathfrak{a}_F)^{K \cap M_F} \cap U(\mathfrak{m}_F + \mathfrak{a}_F)(\mathfrak{z} \cap \mathfrak{m}_F)_c)[x]$$

and

$$(4.21) \qquad u(x) \equiv (x - H_i) v(x) \bmod (\overline{\mathfrak{n}}_{F,c} U(\mathfrak{g}) + U(\mathfrak{g})\mathfrak{z}_c)[x] .$$

The map

$$U(\mathfrak{g})[x] \ni \sum_{j=0}^{N} a_j x^j \longrightarrow \sum_{j=0}^{N} H_i^j a_j \in U(\mathfrak{g})$$

takes $(\overline{\mathfrak{n}}_{F,c} U(\mathfrak{g}) + U(\mathfrak{g})\mathfrak{z}_c)[x]$ to $\overline{\mathfrak{n}}_{F,c} U(\mathfrak{g}) + U(\mathfrak{g})\mathfrak{z}_c$ since $[H_i, \overline{\mathfrak{n}}_F] = \overline{\mathfrak{n}}_F$. Applying this map to (4.21) we get (4.20).

Let $u_j' \in U(\overline{\mathfrak{n}} + \mathfrak{a})$ be given by $u_j - u_j' \in U(\mathfrak{g})\mathfrak{z}_c$, then

$$(4.22) \qquad H_i^m + \Sigma_{j=1}^m H_i^{m-j} u_j' \in \overline{\mathfrak{n}}_{F,c} U(\overline{\mathfrak{n}} + \mathfrak{a}) .$$

Put $D_j = (-1)^{m-j}\Gamma(u_j)$ and let Q_i by given by (4.19). Then, by (4.17), Q_i has a coordinate expression of the form

$$(4.23) \qquad -\Sigma_{j=1}^{m} \left(t_i \frac{\partial}{\partial t_i}\right)^{m-j} \chi_\lambda(D_j) - \left(-t_i \frac{\partial}{\partial t_i}\right)^{m} + t_i P(x,t,\frac{\partial}{\partial x}, t\frac{\partial}{\partial t}) \quad .$$

Moreover, since u_j is of order $\leq j$, $H_i^m + \Sigma_{j=1}^m H_i^{m-j} u_j'$ is of order $\leq m$, whence P is of degree $\leq m$.

Thus Q_i has regular singularities in the weak sense along $\varphi_g(\bar{N} \times \{t_i = 0\})$, and its indicial polynomial is

$$-\Sigma_{j=1}^{m} s^{m-j} \chi_\lambda(D_j) - (-s)^m = -\Sigma_{j=0}^{m} (-s)^{m-j} \gamma(u_j)(\lambda)$$

$$= -\Sigma_{j=0}^{m} (-s)^{m-j} p_j(\lambda) = -\prod_{j=1}^{m} (-s + \rho(H_i) - \lambda(w_j^{-1} H_i))$$

with the roots $(\rho - w_j\lambda)(H_i)$ $(j = 1,\ldots,m)$. □

Remark 4.3.3 Let $F \subset \Delta$. From (4.23) it follows that the system of differential operators $\{Q_i \mid \alpha_i \notin F\}$ has regular singularities in the weak sense along the walls given by $\varphi_g(\bar{N} \times \{t_i = 0\})$ $(\alpha_i \notin F)$ with the edge $\varphi_g(\bar{N} \times \{t_i = 0 \mid \alpha_i \notin F\})$. Let $p = |\Delta \setminus F|$, then the characteristic exponents $s_{\bar{\sigma}} \in \mathbb{C}^p$ are determined by elements $\bar{\sigma} = (\sigma_i)_{\alpha_i \notin F}$ in the product of all $W(H_i)\setminus W$ $(\alpha_i \notin F)$, i.e., $\sigma_i \in W(H_i)\setminus W$ and

$$(4.24) \qquad s_{\bar{\sigma}} = (\langle \rho - \sigma_i \lambda, H_i \rangle)_{\alpha_i \notin F} \in \mathbb{C}^p \quad .$$

Example 4.3.4 Consider once more $X = SL(2,\mathbb{R})/SO(2)$. Since $n = 1$, $\mathbb{D}(G/K)$ has one generator and this is the Laplace-Beltrami operator. It can be seen that in the coordinates

$$X_+ \ni x + it \longrightarrow \begin{pmatrix} x & t \\ 0 & 1 \end{pmatrix} K \in X$$

on X , Δ_L is (up to a constant factor) given by

$$(4.25) \qquad \Delta_L = t^2\left(\frac{\partial^2}{\partial t^2} + \frac{\partial^2}{\partial x^2}\right)$$

(for instance, one can check by brute force that $t^2\left(\frac{\partial^2}{\partial t^2}+\frac{\partial^2}{\partial x^2}\right)$ is invariant under the transformations

$$\begin{pmatrix} a & b \\ c & d \end{pmatrix}(x+it) = \frac{a(x+it)+b}{c(x+it)+d}\ ) \quad .$$

Obviously Δ_L can be continued analytically to a differential operator on $\tilde{X} \simeq \mathbb{C}\mathbb{P}^1$ with regular singularities along X_0. In fact $\tilde{\Delta}_L - \lambda$ is the operator we treated as an example in Section 2.4.

4.4 Notes

The material of this chapter is almost entirely due to T. Oshima [a]. The proof of Theorem 4.1.5 given here follows Oshima's rather close (the reader of [a] (and of Oshima and Sekiguchi [a]) should be aware of the difference between $A(F)$ and A^F, in our notation).

Other compactifications of a similar nature were constructed by I. Satake [a] and H. Furstenberg [a] (see also Moore [a] and Korányi [c]). In fact, the closure $\bar{X}$ of X in $\tilde{X}$ is identical to the maximal Satake-Furstenberg compactification, since it meets Satake's axioms (Korányi [c] p. 349). In Kashiwara et al. [a] a different realization was constructed to solve Helgason's conjecture, but Oshima's construction is more satisfying since it carries a global G-action.

Lemma 4.1.6 is similar to Kashiwara et al. [a] Lemma 4.2. Proposition 4.2.1, Theorem 4.3.1 and their proofs are from Oshima [a]. Proposition 4.3.2 is given in Kashiwara et al. [a] for the realization constructed there. Our proof follows that (a missing ρ has been corrected).

5. Boundary values and Poisson integral representations

Consider the open disk $D = \{|z| < 1\}$ in $\mathbb{C}$ with the boundary $T = \{|z| = 1\}$. The classical Poisson kernel is defined by

$$P(z,t) = \frac{1-|z|^2}{|t-z|^2} \tag{5.1}$$

for $z \in D$, $t \in T$, and the Poisson transform $\mathcal{P}f$ on D of a function f on T is given by

$$\mathcal{P}f(z) = \int_T f(t)\, P(z,t)dt \tag{5.2}$$

As we mentioned in Section 1.1, the Poisson transformation establishes a bijection from hyperfunctions on T to harmonic functions on D, that is, functions annihilated by the operator $\frac{\partial^2}{\partial x^2} + \frac{\partial^2}{\partial y^2}$.

In this chapter we will discuss the much more delicate analog of this statement for the operators

$$(1-x^2-y^2)^2\left(\frac{\partial^2}{\partial x^2} + \frac{\partial^2}{\partial y^2}\right) - s(s-1), \tag{5.3}$$

where $s \in \mathbb{C}$, and their counterparts in the general setting of Riemannian symmetric spaces. That is to say, when D is transformed to the upper half plane, the operator (5.3) is transformed into the operator $\Delta_L - s(s-1)$, where Δ_L is given by (4.25).

For $s = 0$ or $s = 1$ the functions on D annihilated by (5.3) are precisely the harmonic functions.

Our purpose is thus to represent every function u on G/K, which is a joint eigenfunction for $\mathbb{D}(G/K)$, by the "Poisson integral" of a hyperfunction f on the boundary $G/P \cong K/M$.

The way we construct from u the hyperfunction f on K/M is by employing the theory of Chapter 2. From Theorem 4.3.1 we know that the system $\mathcal{M}_\lambda$ on G/K has regular singularities at K/M, and hence u has boundary values in $\mathcal{B}(K/M)$. One of these will be f.

5.1 Poisson transformations

Let G/K be a Riemannian symmetric space of the noncompact type, with notation as in Section 3.1. For each $\lambda \in \mathfrak{a}_c^*$ we define the Poisson kernel by

$$P_\lambda(x,k) = \exp\langle -\lambda-\rho, H(x^{-1}k)\rangle$$

for $x \in G/K$ and $k \in K/M$. For a hyperfunction f on K/M we then define its Poisson integral on G/K by

$$\mathcal{P}_\lambda f(x) = \int_{K/M} f(k)P_\lambda(x,k)dk \tag{5.4}$$

Notice that since K/M is compact, the integral makes sense for each $x \in G/K$. Since for fixed k $P_\lambda(x,k)$ is real analytic in x it follows that $\mathcal{P}_\lambda f$ is a real analytic function on G/K. The mapping $\mathcal{P}_\lambda : \mathcal{B}(K/M) \longrightarrow \mathcal{A}(G/K)$ is called the Poisson transformation.

Example. Let G/K be the Riemannian symmetric space $SL(2,\mathbb{R})/SO(2)$. Instead of identifying it with the upper half plane as we did in Chapter 4, we realize it on the disk D. On D the group

$$SU(1,1) = \left\{ \begin{pmatrix} \alpha & \beta \\ \bar\beta & \bar\alpha \end{pmatrix} \;\middle|\; \alpha, \beta \in \mathbb{C},\ |\alpha|^2 - |\beta|^2 = 1 \right\},$$

isomorphic to $SL(2,\mathbb{R})$, acts transitively by

$$\begin{pmatrix} \alpha & \beta \\ \bar\beta & \bar\alpha \end{pmatrix} \cdot z = \frac{\alpha z + \beta}{\bar\beta z + \bar\alpha} .$$

The isotropy group at 0 is

$$K = \left\{ \begin{pmatrix} \alpha & 0 \\ 0 & \bar\alpha \end{pmatrix} \;\middle|\; |\alpha| = 1 \right\} .$$

We take the Iwasawa decomposition of G given by

$$A = \left\{ \begin{pmatrix} \cosh t & \sinh t \\ \sinh t & \cosh t \end{pmatrix} \;\middle|\; t \in \mathbb{R} \right\} \text{ and } N = \left\{ \begin{pmatrix} 1+ix & -ix \\ ix & 1-ix \end{pmatrix} \;\middle|\; x \in \mathbb{R} \right\} .$$

Let $s \in \mathbb{C}$ and define $\lambda \in \mathfrak{a}_c^*$ by

$$\left\langle \lambda, \begin{pmatrix} 0 & 1 \\ 1 & 0 \end{pmatrix} \right\rangle = 2s-1$$

then by easy computations it follows that

$$P_\lambda(x,t) = \left(\frac{1-|x|^2}{|t-x|^2} \right)^s$$

for $x \in D \cong G/K$ and $t \in T \cong K/M$, where the isomorphism $K/M \xrightarrow{\sim} T$ is given by

$$\begin{pmatrix} \alpha & 0 \\ 0 & \bar{\alpha} \end{pmatrix} M \longrightarrow \alpha^2 \in T \qquad .$$

Thus the Poisson kernel is the classical kernel (5.1), raised to the power s. □

We will now discuss some properties of the Poisson transformation. First, it is convenient to reformulate (5.4).

For $\lambda \in \mathfrak{a}_c^*$ we denote by $\mathcal{B}(G/P;L_\lambda)$ the space of hyperfunctions f on G satisfying

$$(5.5) \qquad f(gman) = a^{\lambda-\rho} f(g)$$

for all $g \in G$, $m \in M$, $a \in A$ and $n \in N$. As before, we use the notation that $a^\nu = \exp\langle \nu, H(a)\rangle$ for $\nu \in \mathfrak{a}_c^*$, $a \in A$. Then $\mathcal{B}(G/P;L_\lambda)$ is canonically identified with the space of hyperfunction valued sections of the line bundle L_λ on G/P associated with the character σ_λ on P given by

$$\sigma_\lambda(man) = a^{\rho-\lambda} \quad .$$

By the Iwasawa decomposition, restriction from G to K defines an isomorphism of $\mathcal{B}(G/P;L_\lambda)$ with $\mathcal{B}(K/M)$. Via this isomorphism we can define the Poisson integral of elements in $\mathcal{B}(G/P;L_\lambda)$, and we have

Lemma 5.1.1 *For* $f \in \mathcal{B}(G/P;L_\lambda)$ *the Poisson integral* (5.4) *of* f *is given by*

$$(5.6) \qquad \mathcal{P}_\lambda f(gK) = \int_K f(gk)dk$$

for $g \in G$.

Proof: From the change of variables in (3.6) we get

$$\int_K f(gk)dk = \int_K f(g\kappa(g^{-1}k)) \exp\langle -2\rho, H(g^{-1}k)\rangle dk$$

Since $f \in \mathcal{B}(G/P;L_\lambda)$ we have $f(x) = f(\kappa(x)) \exp\langle \lambda-\rho, H(x)\rangle$, and it is easily seen that $\kappa(g\kappa(g^{-1}k)) = k$ and $H(g\kappa(g^{-1}k)) = -H(g^{-1}k)$. Inserted this gives

$$\int_K f(gk)dk = \int_K f(k) \exp\langle -\lambda-\rho, H(g^{-1}k)\rangle dk$$

as claimed in (5.6). □

In particular we see that $\mathcal{P}_\lambda$ is a G-map from $\mathcal{B}(G/P;L_\lambda)$ to functions on G/K, when G acts on both spaces by translation.

Let $\mathcal{A}(G/K;\mathcal{M}_\lambda)$ denote the space of hyperfunctions φ on G/K satisfying $D\varphi = \chi_\lambda(D)\varphi$ for all $D \in \mathbb{D}(G/K)$ $(\lambda \in \mathfrak{a}_c^*)$. Since $\mathbb{D}(G/K)$ contains the Laplacian operator Δ_L which is elliptic, $\mathcal{A}(G/K;\mathcal{M}_\lambda)$ consists of analytic functions only (cf. Theorem 1.5.5).

Proposition 5.1.2 The Poisson transformation $\mathcal{P}_\lambda$ maps $\mathcal{B}(G/P;L_\lambda)$ into $\mathcal{A}(G/K;\mathcal{M}_\lambda)$ for all $\lambda \in \mathfrak{a}_c^*$.

Proof: Let $u \in U(\mathfrak{g})^K$ and let $D = \Gamma(u)$. Then, since u commutes with K we have from Lemma 5.1.1

$$D\mathcal{P}_\lambda f(gK) = \int_K (u_r f)(gk)dk$$

where u_r denotes u acting from the right on f. From (3.15) and (5.5) it immediately follows that

$$D\mathcal{P}_\lambda f(gK) = \delta(u)(\lambda-\rho)\mathcal{P}_\lambda f(gK) = \chi_\lambda(D)\mathcal{P}_\lambda f(gK) \ . \quad \square$$

Let $\delta \in \mathcal{B}(K/M)$ denote the Dirac measure supported at the origin, then from (5.4) we have

$$P_\lambda(x,k) = \mathcal{P}_\lambda\delta(k^{-1}x)$$

and hence it follows that for each $k \in K/M$ the Poisson kernel $P_\lambda(x,k)$ as a function of $x \in G/K$ is an eigenfunction for each $D \in \mathbb{D}(G/K)$ with eigenvalue $\chi_\lambda(D)$.

We will now, for a certain range of λ's, prove that a hyperfunction on K/M is the "radial limit" of its Poisson transforms (at least in a weak sense). For this we need some information on a very crucial integral. Let

$$\mathfrak{a}_+^* = \{\lambda \in \mathfrak{a}^* \mid \langle\lambda,\alpha\rangle > 0 \ \forall\alpha \in \Sigma^+\} .$$

<u>Proposition 5.1.3</u> <u>For each</u> $\lambda \in \mathfrak{a}_c^*$ <u>such that</u> $\mathrm{Re}\,\lambda \in \mathfrak{a}_+^*$ <u>the integral</u>

$$c_\lambda = \int_{\overline{N}} \exp\langle -\lambda-\rho, H(\overline{n})\rangle \, d\overline{n}$$

<u>converges absolutely to a constant</u> $c_\lambda \neq 0$. <u>Moreover</u> c_λ <u>is given by the expression</u>

$$(5.7) \qquad c_\lambda = C_o \prod_{\alpha \in \Sigma_o^+} \frac{\Gamma\left(\frac{\langle\lambda,\alpha\rangle}{\langle\alpha,\alpha\rangle}\right) 2^{-\frac{\langle\lambda,\alpha\rangle}{\langle\alpha,\alpha\rangle}}}{\Gamma\left(\frac{1}{2}\left(\frac{1}{2}m(\alpha)+1+\frac{\langle\lambda,\alpha\rangle}{\langle\alpha,\alpha\rangle}\right)\right)\Gamma\left(\frac{1}{2}\left(\frac{1}{2}m(\alpha)+m(2\alpha)+\frac{\langle\lambda,\alpha\rangle}{\langle\alpha,\alpha\rangle}\right)\right)}$$

<u>where</u> $\Sigma_o^+ = \Sigma^+ \setminus 2\Sigma^+$ <u>is the set of indivisible positive roots, and</u> C_o <u>is a constant independent of</u> λ.

<u>Proof</u>: (Harish-Chandra [c], Gindikin and Karpelevič [a]) See Helgason [n] Chapter 4, Theorem 6.14. □

In the following theorem and corollary the notation $a \to \infty$ means that $a \in A$ and $a^\alpha \to \infty$ for all $\alpha \in \Delta$ (the set of simple roots for Σ^+).

<u>Theorem 5.1.4</u> <u>Let</u> $\lambda \in \mathfrak{a}_c^*$ <u>be such that</u> $\mathrm{Re}\,\lambda \in \mathfrak{a}_+^*$ <u>and let</u> $f \in \mathcal{B}(G/P;L_\lambda)$ <u>be continuous. Then</u>

$$(5.8) \qquad f(x) = c_\lambda^{-1} \lim_{a \to \infty} a^{\rho-\lambda} \mathcal{P}_\lambda f(x a K)$$

<u>for every</u> $x \in G$, <u>with uniform convergence on compact sets.</u>

<u>Proof</u>: We transform the integral (5.6) into an integral over $\overline{N}$ using (3.7):

$$\mathcal{P}_\lambda f(xaK) = \int_{\overline{N}} f(xa\kappa(\overline{n}))\exp\langle -2\rho, H(\overline{n})\rangle d\overline{n}$$

$$= \int_{\overline{N}} f(xa\overline{n})\exp\langle -\lambda-\rho, H(\overline{n})\rangle d\overline{n}$$

$$= a^{\lambda-\rho}\int_{\overline{N}} f(xa\overline{n}a^{-1})\exp\langle -\lambda-\rho, H(\overline{n})\rangle d\overline{n}$$

Now $a\overline{n}a^{-1} \to e$ as $a \to \infty$, and a reversal, if justified, of the order of the limit and the integration will thus prove the theorem.

To justify this step by the dominated convergence theorem, it suffices to prove that the functions

$$\varphi_a(\overline{n}) = \exp\langle \mathrm{Re}\,\lambda-\rho, H(a\overline{n}a^{-1})\rangle \exp\langle -\mathrm{Re}\,\lambda-\rho, H(\overline{n})\rangle$$

are uniformly in $L^1(\overline{N})$ as $a \to \infty$, since

$$|f(xa\overline{n}a^{-1})| \leq \exp\langle \mathrm{Re}\,\lambda-\rho, H(a\overline{n}a^{-1})\rangle \sup_{k\in K}|f(xk)| \quad .$$

Let ${}^+\mathfrak{a} = \{H \in \mathfrak{a} \mid \langle \nu, H\rangle \geq 0,\ \forall \nu \in \mathfrak{a}_+^*\}$, then ${}^+\mathfrak{a}$ is the cone dual to $\mathfrak{a}^+$. We need the following facts (cf. Helgason [n] Chapter 4, Corollary 6.6 - see also Lemma 6.1.6 below)

(5.9) $\qquad H(\overline{n}) \in {}^+\mathfrak{a}$ and $H(\overline{n}) - H(a\overline{n}a^{-1}) \in {}^+\mathfrak{a}$

for all $\overline{n} \in \overline{N}$, $a \in A^+$.

Choose ε such that $0 < \varepsilon < 1$ and $\rho-\varepsilon\mathrm{Re}\,\lambda \in \mathfrak{a}_+^*$. Then from the first part of (5.9) we get

$$\exp\langle -\rho, H(a\overline{n}a^{-1})\rangle \leq \exp - \varepsilon\langle \mathrm{Re}\,\lambda, H(a\overline{n}a^{-1})\rangle$$

and from the second part of (5.9)

$$\exp(1-\varepsilon)\langle \mathrm{Re}\,\lambda, H(a\overline{n}a^{-1})\rangle \leq \exp(1-\varepsilon)\langle \mathrm{Re}\,\lambda, H(\overline{n})\rangle$$

from which it follows that

$$\exp\langle \mathrm{Re}\,\lambda-\rho, H(a\overline{n}a^{-1})\rangle \leq \exp(1-\varepsilon)\langle \mathrm{Re}\,\lambda, H(\overline{n})\rangle$$

and hence φ_a is dominated by

$$\exp\langle -\varepsilon\mathrm{Re}\,\lambda-\rho, H(\overline{n})\rangle$$

which is integrable by Proposition 5.1.3. □

Remark. The proof actually shows that (5.8) holds whenever $f \in \mathcal{B}(G/P;L_\lambda)$, $f|_K \in L^\infty(K)$ and x is a point of continuity for f.

Corollary 5.1.5 Let $\lambda \in \mathfrak{a}_c^*$ be such that $\operatorname{Re}\lambda \in \mathfrak{a}_+^*$ and let $f \in \mathcal{B}(K/M)$. Then for each analytic function $\varphi \in \mathcal{A}(K)$

$$\int_K f(kM)\varphi(k)\,dk = c_\lambda^{-1} \lim_{a \to \infty} a^{\rho-\lambda} \int_K \mathcal{P}_\lambda f(k\,a\,K)\varphi(k)\,dk$$

i.e., $c_\lambda^{-1} a^{\rho-\lambda}\mathcal{P}_\lambda f(\cdot\, a\,K)$ converges to f in the weak sense as analytic functionals.

Proof: Define

$$\varphi * f(k) = \int_K \varphi(kl^{-1}) f(lM)\,dl$$

for $k \in K$. Since φ is analytic, so is $\varphi * f$. We thus have $\varphi * f \in \mathcal{A}(K/M)$. By Theorem 5.1.4

$$\varphi * f(e) = c_\lambda^{-1} \lim_{a \to \infty} a^{\rho-\lambda} \mathcal{P}_\lambda(\varphi * f)(a\,K) \quad .$$

However, for $x \in G/K$ we have

$$\begin{aligned}
\mathcal{P}_\lambda(\varphi * f)(x) &= \int_K \int_K \varphi(kl^{-1}) f(lM) P_\lambda(x,k)\,dl\,dk \\
&= \int_K \int_K \varphi(k) f(lM) P_\lambda(x,kl)\,dk\,dl \\
&= \int_K \varphi(k) \int_K f(lM) P_\lambda(k^{-1}x,l)\,dl\,dk \\
&= \int_K \varphi(k)\, \mathcal{P}_\lambda f(k^{-1}x)\,dk
\end{aligned}$$

from which the corollary follows. □

Corollary 5.1.6 For all $\lambda \in \mathfrak{a}_c^*$ such that $\operatorname{Re}\lambda \in \mathfrak{a}_+^*$ the Poisson transformation $\mathcal{P}_\lambda : \mathcal{B}(K/M) \longrightarrow \mathcal{A}(G/K)$ is injective.

Proof: Follows immediately from the preceding corollary. □

5.2 Boundary value maps

Let $\lambda \in \mathfrak{a}_c^*$ and let $u \in \mathcal{A}(G/K; \mathcal{M}_\lambda)$. By Theorem 4.3.1, Proposition 4.3.2 and Section 2.5 we can define boundary value of u , provided the difference of any two characteristic exponents does not belong to $\mathbb{Z}^n$. Since the characteristic exponents are $s_{w,j} = \langle \rho - w\lambda, H_j \rangle$ $(j = 1,\dots,n\,;\, w \in W)$ this happens if and only if λ satisfies the following

Assumption (A):

For each $w \in W \setminus \{e\}$ there exists $i \in \{1,\dots,n\}$ such that $\langle w\lambda - \lambda, H_i \rangle \notin \mathbb{Z}$.

Recall from Section 2.5 the line bundle

$$\mathcal{L}_w = \bigotimes_{j=1}^{n} \left(T^*_{N_j} \tilde{X} \right)^{\otimes s_{w,j}}$$

on O_0 , where $O_0 = N_1 \cap \dots \cap N_n \simeq G/P$ is the edge. Since G acts on $\tilde{X}$ and preserves each N_j there is a natural action of G on $\mathcal{L}_w$. The action of $y \in G$ on the sections of $\mathcal{L}_w$ can be described as follows. Let $g \in G$ and consider the coordinate systems $\varphi_g(\bar{n}, t)$ and $\varphi_{g'}(\bar{n}', t')$ where $g' = yg$. Then the action of y on $\mathcal{L}_w$ is given by

$$y \cdot (dt)^{s_w} = (dt')^{s_w} \qquad (5.10)$$

Since G acts on $\mathcal{L}_w$, it acts on the space $\mathcal{B}(O_0 ; \mathcal{L}_w)$ of hyperfunction sections of $\mathcal{L}_w$.

Assume (A), then we have the boundary value map

$$\beta : \mathcal{A}(G/K ; \mathcal{M}_\lambda) \longrightarrow \bigoplus_{w \in W} \mathcal{B}(O_0 ; \mathcal{L}_w) .$$

From (5.10) it follows that β is a G-map, by the independence of β on local coordinates.

We shall now give a more transparent interpretation of $\mathcal{L}_w$. In order to avoid later repetition we give a more general statement. Thus let $F \subset \Delta$ and $\mu \in (\mathfrak{a}_F)^*_c$ and let $\mathcal{L}^F_\mu$ be the line bundle

$$\mathcal{L}^F_\mu = \bigotimes_{\alpha_j \notin F} \left(T^*_{N_j}\widetilde{X}\right)^{\otimes<\rho-\mu, H_j>}$$

on $\bigcap_{\alpha_j \notin F} N_j$. For the coordinate patch $\varphi_g(\bar{n}, t)$ let $dt^{\rho-\mu}$ denote $\prod_{\alpha_j \notin F} dt_j^{<\rho-\mu, H_j>}$. Similarly to (5.10) there is a natural action of G on $\mathcal{L}^F_\mu$ given by

$$(5.11) \qquad y \cdot (dt)^{\rho-\mu} = (dt')^{\rho-\mu}$$

in coordinates $\varphi_g(\bar{n}, t)$ and $\varphi_{g'}(\bar{n}, t)$ respectively, where $g' = yg$.

Let O be an open orbit of $\bigcap_{\alpha_j \notin F} N_j$, then $O \simeq G/B_F$.

<u>Lemma 5.2.1</u> <u>$\mathcal{L}^G_\mu$ on O is isomorphic to L^F_μ on G/B_F by an isomorphism equivariant for G.</u>

<u>Proof</u>: For simplicity assume $F = \{\alpha_{k+1}, \ldots, \alpha_n\}$ $(0 < k \leq n)$. Write the element t of $\mathbb{R}^n$ as $t = (t_0, t_1) \in \mathbb{R}^k \times \mathbb{R}^{n-k}$.

Let $g, g' \in G$ and consider the coordinates $\varphi_g(\bar{n}, t)$ and $\varphi_{g'}(\bar{n}', t')$ near the point $q \in O$. From Lemma 4.1.6 it follows that the relation between dt_j and dt'_j at q is

$$dt'_j = \exp<-\alpha_j, H_{g'^{-1}g}(\bar{n}, t)> dt_j$$

for $j \leq k$ where $q = \varphi_g(\bar{n}, t) \in O$. It follows that

$$(dt')^{\rho-\mu} = \exp\left[-\sum_{j=1}^{n-k} <\rho-\mu, H_j><\alpha_j, H_{g'^{-1}g}(\bar{n}, t)>\right] dt^{\rho-\mu} .$$

By definition of $H_{g'^{-1}g}$ we have

$$g'^{-1}g\,\bar{n}a_t \in \bar{N} \exp H_{g'^{-1}g}(\bar{n}, t)(M_F \cap K)\, N_F .$$

Writing $q = \varphi_g(\bar{n},t) = \varphi_{g'}(\bar{n},t')$ we have $a_{t'}^{-1}\,\bar{n}'^{-1}\,g'^{-1}\,g\bar{n}a_t \in B_F$ and so it follows that

$$H_{g'^{-1}g}(\bar{n},t) = \log a_{t'} + H(a_{t'}^{-1}\,\bar{n}'^{-1}g'^{-1}\,g\bar{n}a_t) \quad .$$

Since $\alpha_j(\log a_{t'}) = 0$ for $j \leq n-k$ we get

$$(5.12) \qquad (dt')^{\rho-\mu} = \sigma_\mu^F(a_{t'}^{-1}\,\bar{n}'^{-1}g'^{-1}g\bar{n}a_t)^{-1}(dt)^{\rho-\mu} \quad .$$

On the other hand, by definition L_μ^F is $G \times \mathbb{C}$ modulo the equivalence relation $(xb,c) \sim (x,\sigma_\mu^F(b)c)$ $(x \in G\ ,\ b \in B_F\ ,\ c \in \mathbb{C})$. Therefore:

$$(5.13) \qquad (g'\bar{n}'a_{t'}\,,\,1) \sim \sigma_\mu^F(a_{t'}^{-1}\,\bar{n}'^{-1}g'^{-1}g\bar{n}a_t)^{-1}(g\bar{n}a_t,\,1) \; .$$

From (5.12) and (5.13) the isomorphism between $\mathcal{L}_\mu^F$ and L_μ^F follows. By (5.11) it is equivariant for G . □

Using the lemma with $E = \emptyset$ we get:

<u>Corollary 5.2.2</u> <u>Under Assumption (A), the boundary value map results in a G-map</u>

$$\beta = \bigoplus_{w\in W} \beta_{w\lambda} : \mathcal{A}(G/K;\mathcal{M}_\lambda) \longrightarrow \bigoplus_{w\in W} \mathcal{B}(G/P;L_{w\lambda}) \quad .$$

For later reference we notice that if Z is a real analytic manifold and we denote by $\mathcal{B}(Z\times G/K;\mathcal{M}_\lambda)$ the space of hyperfunctions on $Z\times G/K$ which solves $\mathcal{M}_\lambda$ in the second variable, then we can define boundary value maps

$$(5.14) \qquad \beta_{w\lambda} : \mathcal{B}(Z\times G/K;\mathcal{M}_\lambda) \longrightarrow \mathcal{B}(Z\times G/P;L_{w\lambda})$$

where $\mathcal{B}(Z\times G/P;L_{w\lambda})$ denotes the space of hyperfunctions f on $Z\times G$ satisfying

$$f(z,gp) = \sigma_{w\lambda}^{-1}(p)f(z,g)$$

for $z \in Z$, $g \in G$, and $p \in P$. This follows trivially from what has been done above, since $z \in Z$ does not enter the differential equations at all.

Finally in this section we consider Assumption (A) closer.

<u>Lemma 5.2.3</u> <u>Let</u> $\lambda \in \mathfrak{a}_c^*$. <u>Then</u> λ <u>satisfies (A) if and only if</u> $\frac{2<\lambda,\alpha>}{<\alpha,\alpha>} \notin \mathbb{Z}$ <u>for all</u> $\alpha \in \Sigma$.

<u>Proof</u>: If λ satisfies (A) and $\alpha \in \Sigma$, then $<w_\alpha\lambda-\lambda, H_i> \notin \mathbb{Z}$ for some i where w_α denotes reflection in α, i.e., $w_\alpha\lambda = \lambda + \frac{2<\lambda,\alpha>}{<\alpha,\alpha>}\alpha$. Since $\alpha(H_i)$ is an integer, $\frac{2<\lambda,\alpha>}{<\alpha,\alpha>}$ cannot be an integer.

Assume conversely that λ does not satisfy (A), i.e., $w\lambda - \lambda \in \mathrm{Span}_{\mathbb{Z}}\Sigma$ for some Weyl group element $w \neq e$. Write $\lambda = \xi + \sqrt{-1}\zeta$ with ξ and ζ in $\mathfrak{a}^*$ and define for $\mu \in \mathfrak{a}^*$: $\sigma\mu = w\mu + \lambda - w\lambda = w\mu + \xi - w\xi$. Then σ belongs to the affine Weyl group of Σ (Bourbaki [a] VI §2 n° 2.1) and $\sigma(\xi + c\zeta) = \xi + c\zeta$ for all $c \in \mathbb{R}$. By loc. cit. V §3 n° 3.3 Proposition 2 there exists $\alpha \in \Sigma$ such that $\frac{2<\xi+c\zeta,\alpha>}{<\alpha,\alpha>} \in \mathbb{Z}$ for all $c \in \mathbb{R}$ whence $\frac{2<\xi,\alpha>}{<\alpha,\alpha>} \in \mathbb{Z}$ and $<\zeta,\alpha> = 0$, i.e., $\frac{2<\lambda,\alpha>}{<\alpha,\alpha>} \in \mathbb{Z}$. □

5.3 <u>Spherical functions and their asymptotics</u>

For $\lambda \in \mathfrak{a}_c^*$ let ϕ_λ denote the Poisson integral on G/K of the constant function 1 on K/M, i.e.,

$$\phi_\lambda(g) = \int_K \exp<-\lambda-\rho, H(g^{-1}k)> dk$$

or, by Lemma 5.1.1, equivalently

$$\phi_\lambda(g) = \int_K \exp<\lambda-\rho, H(gk)> dk .$$

The functions ϕ_λ are called the (elementary) <u>spherical functions</u> for G.

By an elementary standard argument (see Helgason [n] Chapter 1, Corollary 2.3) ϕ_λ is uniquely determined in $\mathcal{A}(G/K;\mathcal{M}_\lambda)$ by being K-invariant from the left and taking the value 1 at the identity.

<u>Lemma 5.3.1</u> <u>Let</u> $\lambda, \mu \in \mathfrak{a}_c^*$. <u>Then</u> $\phi_\lambda = \phi_\mu$ <u>if and only if</u> $\mu = s\lambda$ <u>for some</u> $s \in W$.

<u>Proof</u>: If $\phi_\lambda = \phi_\mu$ then $\chi_\lambda(D) = \chi_\mu(D)$ for all $D \in \mathbb{D}(G/K)$, whence $\mu = s\lambda$ for some $w \in W$. The converse statement similarly follows (and is also proved in Proposition 7.3.5). □

Assume $\lambda \in \mathfrak{a}_c^*$ satisfies Assumption (A). For each $w \in W$ the map $\beta_{w\lambda}$ is a G-map, and hence $\beta_{w\lambda}\,\phi_\lambda$ is invariant under K from the left. From (5.5) we conclude that if $c(w,\lambda) = \beta_{w\lambda}\,\phi_\lambda(e) \in \mathbb{C}$ then

$$\beta_{w\lambda}\,\phi_\lambda(x) = c(w,\lambda)\,\exp\langle w\lambda-\rho, H(x)\rangle \quad . \tag{5.15}$$

From the preceding lemma we get that $c(ws^{-1}, s\lambda) = c(w,\lambda)$, and letting $c(\lambda) = c(e,\lambda) = \beta_\lambda\,\phi_\lambda(e)$ we therefore have $c(w,\lambda) = c(w\lambda)$.

In particular, from (5.15) we get that the boundary values of ϕ_λ are analytic functions, so that we can apply Theorem 2.5.6. Recall that for each point t of the set $\mathbb{R}^n_+ = \{t \in \mathbb{R}^n \mid t_j > 0\,,\ j = 1,\ldots,n\}$ we have defined

$$a_t = \exp(-\,\Sigma_{j=1}^n\,(\log t_j)H_j) \in A \quad .$$

Let $\mathfrak{a}_c^{*\prime}$ denote the set of those $\lambda \in \mathfrak{a}_c^*$ that satisfy Assumption (A).

<u>Theorem 5.3.2</u> <u>For each</u> $w \in W$ <u>there exists a function</u> $h_w(\lambda,t)$, <u>defined on a neighborhood</u> Ω <u>of</u> $\mathfrak{a}_c^{*\prime} \times \{0\}$ <u>in</u> $\mathfrak{a}_c^{*\prime} \times \mathbb{R}^n$, <u>holomorphic in</u> λ <u>and real analytic in</u> t , <u>and with</u> $h_w(\lambda,0) = c(w\lambda)$, <u>such that</u>

$$\phi_\lambda(a_t K) = \Sigma_{w\in W}\, h_w(\lambda,t)\,a_t^{w\lambda-\rho} \tag{5.16}$$

<u>for all</u> $(\lambda,t) \in \Omega$ <u>with</u> $t \in \mathbb{R}^n_+$.

<u>Proof</u>: Consider the local coordinates on $\widetilde{X}$ given by $\varphi_e : \overline{N} \times \mathbb{R}^n \longrightarrow \widetilde{X}$. It follows from Theorem 2.5.6 and Section 2.5.2 that

$$\phi_\lambda(\bar{n}a_tK) = \Sigma_{w\in W}\ \varphi_w(\lambda,\bar{n},t) \prod_{j=1}^{n} t_j^{<\rho-w\lambda,H_j>}$$

for some functions $\varphi_w(\lambda,\bar{n},t)$ analytic in $(\bar{n},t)$ in a neighborhood of $\bar{N}\times\{0\}$, and holomorphic in $\lambda\in\mathfrak{a}_c^{*\prime}$. Taking $\bar{n}=e$, the theorem follows. □

Later on, in Theorem 6.3.4, we will prove a generalization to this theorem, showing a similar behavior of $\phi(a_tK)$ when only some of the coordinates t_j tend to 0.

In particular, it follows from Theorem 5.3.2 that $c(\lambda)$ depends holomorphically on λ in $\mathfrak{a}_c^{*\prime}$. We will now use (5.16) to identify $c(\lambda)$.

<u>Proposition 5.3.3</u> <u>If</u> $\lambda\in\mathfrak{a}_c^{*\prime}$ <u>and</u> $\mathrm{Re}\,\lambda\in\mathfrak{a}_+^*$ <u>then the following identity holds for</u> $c(\lambda)=\beta_\lambda\,\phi_\lambda(e)$:

$$c(\lambda) = \int_{\bar{N}} \exp<-\lambda-\rho,H(\bar{n})>d\bar{n} \quad .$$

<u>Proof</u>: Choose $H_o\in\mathfrak{a}^+$ and put $h_s=\exp s\,H_o$ for $s\in\mathbb{R}$. By Helgason [j] p. 293, $\mathrm{Re}<w\lambda-\lambda,H_o><0$ for $w\in W$, $w\neq e$. Therefore from Theorem 5.3.2 we get

$$c(\lambda) = \lim_{s\to\infty} h_s^{\rho-\lambda}\,\phi_\lambda(h_sK) \quad .$$

Since ϕ_λ is the Poisson transform of the constant function 1 it follows from Theorem 5.1.4 that this limit also equals c_λ. □

From (5.7) we now have an explicit expression for $c(\lambda)$ in terms of the roots of Σ^+ and their multiplicities, provided $\mathrm{Re}\,\lambda\in\mathfrak{a}_+^*$. However, using the fact (Helgason [j] p. 530) that if 2α and α both are roots then $m(\alpha)$ is even and $m(2\alpha)$ odd, it is easy to see that the right hand side of (5.7) is a never vanishing holomorphic function in λ on

$$\{\lambda\in\mathfrak{a}_c^*\ |\ \frac{2<\lambda,\alpha>}{<\alpha,\alpha>}\notin -\mathbb{Z}_+,\ \ \forall\alpha\in\Sigma^+\} \quad .$$

Denoting this meromorphic extension of c_λ also by c_λ, we get from Proposition 5.3.3 that $c(\lambda)=c_\lambda$ for all $\lambda\in\mathfrak{a}_c^{*\prime}$ by analytic continuation.

5.4 Integral representations

Recall from Lemma 5.2.3 that Assumption (A) for $\lambda \in \mathfrak{a}_c^*$ can be stated as follows:

(A) $$\frac{2<\lambda,\alpha>}{<\alpha,\alpha>} \notin \mathbb{Z} \quad \text{for all} \quad \alpha \in \Sigma .$$

The main theorem of this section can now be stated as follows:

Theorem 5.4.1 *If $\lambda \in \mathfrak{a}_c^*$ satisfies (A) the Poisson transformation $\mathcal{P}_\lambda$ is a bijection of $\mathcal{B}(K/M)$ onto $\mathcal{A}(G/K;\mathcal{M}_\lambda)$. Its inverse is c_λ^{-1} times the boundary value map β_λ.*

Before we prove the theorem, we need a lemma. Let $U \in \mathcal{B}(K \times G/K)$ and suppose U solves $\mathcal{M}_\lambda$ in the second variable. By (5.14), U has boundary values $\beta_{w\lambda}U \in \mathcal{B}(K \times G/P;L_{w\lambda})$. Let

$$v(x) = \int_K U(k,x)dk$$

then $v \in \mathcal{A}(G/K;\mathcal{M}_\lambda)$.

Lemma 5.4.2 $\beta_{w\lambda}v(g) = \int_K \beta_{w\lambda}U(k,g)dk$ *for* $g \in G$, *as an equation of hyperfunctions.*

Proof: Let $\tilde{v} \in \mathcal{B}(\tilde{X})$ be the unique extension of v with support in $\bar{X}$ which solves $\mathcal{M}_\lambda$. Define $\tilde{U} \in \mathcal{B}(K \times \tilde{X})$ similarly for U. By the uniqueness

$$\tilde{v}(\tilde{x}) = \int_K \tilde{U}(k,\tilde{x})dk$$

for $\tilde{x} \in \tilde{X}$. Let $\varphi_w(k,x)$ denote the boundary values of $U(k,x)$ in some given local coordinates (x,t) on $\tilde{X}$. Then with the notation of Chapter 2

$$\begin{aligned} sp(\tilde{v}) &= sp\left(\int_K \tilde{U}(k,\tilde{x})dk\right) \\ &= \int_K sp\,\tilde{U}(k,\tilde{x})dk \\ &= \int_K \Sigma_{w\in W} A_w\, sp(\varphi_w(k,x)\; t_+^{s_{w\lambda}})dk \\ &= \Sigma_{w\in W} A_w\, sp\left(\int_K \varphi_w(k,x)dk\; t_+^{s_{w\lambda}}\right) \end{aligned}$$

and hence $\int_K \varphi_w(k,x)dk$ is the boundary value of v. □

Proof of Theorem 5.4.1: Let $u \in \mathcal{A}(G/K;\mathcal{M}_\lambda)$. Notice that

$$\int_K u(kx)dk = u(e)\phi_\lambda(x)$$

since the left side is K-invariant from both sides. By Section 5.3 it follows that

$$\beta_\lambda\left[\int_K u(k\cdot)dk\right](y) = c_\lambda u(e)$$

for $y \in K$. Taking $U(k,x) = u(kx)$ in the preceding lemma, it follows that

$$\int_K \beta_\lambda u(k)dk = c_\lambda u(e)$$

which shows that $\mathcal{P}_\lambda\,\beta_\lambda u(e) = c_\lambda u(e)$. Since both $\mathcal{P}_\lambda$ and β_λ are G-maps we thus have

$$\mathcal{P}_\lambda\,\beta_\lambda u = c_\lambda u$$

for all $u \in \mathcal{A}(G/K;\mathcal{M}_\lambda)$. Thus $\mathcal{P}_\lambda$ is surjective.

From Corollary 5.1.6 it then follows that if $\operatorname{Re}\lambda \in \mathfrak{a}^*_+$ then $\mathcal{P}_\lambda$ is bijective with $c_\lambda^{-1}\beta_\lambda$ as inverse, i.e., we also have $\beta_\lambda\mathcal{P}_\lambda f = c_\lambda f$ for all $f \in \mathcal{B}(K/M)$. By Section 2.5.2 $\beta_\lambda\mathcal{P}_\lambda f$ is holomorphic in λ, and therefore $\beta_\lambda\mathcal{P}_\lambda f = c_\lambda f$ holds for all $\lambda \in \mathfrak{a}^*_c$, by analytic continuation. □

The restriction that λ satisfies (A) is necessary in order to define the boundary value map β_λ as above. This excludes, however, many interesting cases (e.g., $\lambda=\rho$). Without proof we will now mention a more refined result for the Poisson transformation.

Let e_λ^{-1} denote the denominator in (5.7) of the function c_λ, that is, e_λ^{-1} is the meromorphic function

$$(5.17)\quad e_\lambda^{-1} = \prod_{\alpha\in\Sigma_o^+} \Gamma\left(\frac{1}{2}\left(\frac{1}{2}m(\alpha)+1+\frac{<\lambda,\alpha>}{<\alpha,\alpha>}\right)\right)\Gamma\left(\frac{1}{2}\left(\frac{1}{2}m(\alpha)+m(2\alpha)+\frac{<\lambda,\alpha>}{<\alpha,\alpha>}\right)\right).$$

Theorem 5.4.3 Let $\lambda \in \mathfrak{a}^*_c$ and $\mathcal{P}_\lambda : \mathcal{B}(K/M) \longrightarrow \mathcal{A}(G/K;\mathcal{M}_\lambda)$ the Poisson transformation. Then the following statements are equivalent:

(i) $e_\lambda \neq 0$,

(ii) $\mathcal{P}_\lambda$ is injective,

(iii) $\mathcal{P}_\lambda$ is surjective.

The equivalence of (i) and (ii) is proved in Helgason [g] §6 (cf. [c] p. 94). In Kashiwara et al. [a] §6 it is proved that (i) implies (iii)(and (ii)). As a consequence, Corollary 5.4.4 below holds. From this (or from Helgason [g] Corollary 7.4) it follows that for each K-type δ the spaces of K-finite vectors of type δ in $\mathcal{A}(G/K;\mathcal{M}_\lambda)$ and $\mathcal{B}(K/M)$ have the same finite dimension. From this (iii) $\Longrightarrow$ (ii) easily follows, since $\mathcal{P}_\lambda$ is injective if and only if its restriction to the K-finite vectors is injective.

For each $\lambda \in \mathfrak{a}_c^*$ there exists $w \in W$ such that $\operatorname{Re} w\lambda \in \overline{\mathfrak{a}_+^*}$ which implies $e_{w\lambda} \neq 0$. Thus the theorem has the following

Corollary 5.4.4 Let $u \in \mathcal{A}(G/K)$ be a joint eigenfunction for $\mathbb{D}(G/K)$. Then there exists a $\lambda \in \mathfrak{a}_c^*$ and a hyperfunction $f \in \mathcal{B}(K/M)$ such that

$$u(gK) = \int_{K/M} \exp\langle -\lambda-\rho, H(g^{-1}k)\rangle f(kM)dk .$$

Here λ can be chosen such that $\operatorname{Re}\langle\lambda,\alpha\rangle \geq 0$ for all $\alpha \in \Sigma^+$ and then f is uniquely determined by u and λ.

5.5 Notes and further results

The generalization to Riemannian symmetric spaces of the classical Poisson integral does back to H. Furstenberg [a] (see also Lowdenslager [a] and Hua [a]). The first case considered was that of $\lambda = \rho$, for which the Poisson integrals are harmonic functions on G/K. For general λ, Poisson integrals were studied in Furstenberg [b] and Karpelevič [b]. See also Moore [a]I.

Theorem 5.1.4 is due to S. Helgason ([c] Lemma 1.2). For $\lambda = \rho$ it was proved in Karpelevič [b] (Theorem 18.3.2). For related "Fatou-theorems" see Helgason and Korányi [a], Korányi [a], and Knapp and Williamson [a] for $\lambda = \rho$, and Michelson [a] for extensions to other eigenvalues. For further variations see Weiss [a], Urakawa [a],[b], Korányi [c], Lindahl [a], and Stein [a]. See also the notes to Chapter 6. For rank one, other results are given in Knapp[a], Korányi and Putz [a], Mantero [a], Korányi and Taylor [a] and Cygan [a]. See also Sjögren [b],[c]. Surveys of these "Fatou-theorems" and related results are given in Korányi [b], [f].

Corollary 5.1.6 is also from Helgason [c]. Corollary 5.2.2 and Lemma 5.2.3 are proved in Kashiwara et al. [a].

The theory of spherical functions for Riemannian symmetric spaces is given a thorough treatment in Helgason's books [b] and [n]. The integral formula we use as definition and the asymptotic expansion of Theorem 5.3.2 are due to Harish-Chandra [c]I. The proof given here based on Theorem 2.5.6 is due to T. Oshima and J. Sekiguchi [a]. Proposition 5.3.3 is from Harish-Chandra [c]I.

The statement (Corollary 5.4.4) that all eigenfunctions on G/K are Poisson integrals of analytic functionals on K/M was conjectured by S. Helgason, who proved it in [c] and [d] for the hyperbolic disk (see p. 80). A simple exposition of this is given in [m]. This proof was generalized to G of real rank one (except for some condition on λ) in Helgason [f] (see also Hashizume et al. [b] and Minemura [a], [b], [c]). For K-finite eigenfunctions the representation by Poisson integrals was proved in Helgason [g], where also the equivalence of (i) and (ii) in Theorem 5.4.3 is given. The conjecture (i.e., Corollary 5.4.4) was settled in general by M. Kashiwara, A. Kowata, K. Minemura, K. Okamoto, T. Oshima and M. Tamaka in [a], where Theorem 5.4.1 and Theorem 5.4.3 (i) $\Longrightarrow$ (iii) are proved. Our proof of Theorem 5.4.1 follows theirs, except that it is simplified to avoid use of the results in Appendix I of [a]. Theorem 5.4.3 (iii) $\Longrightarrow$ (ii) was observed by Helgason (cf. [n]).

The eigenfunctions on G/K which are Poisson integrals of distributions are determined in Lewis [a] (rank one) and Oshima and Sekiguchi [a] (in general) (see also Wallach [c]). The condition for the boundary value to be a distribution is a certain growth condition on the eigenfunction.

The eigenfunctions which are Poisson integrals of L^p-functions $(1 < p \leq \infty)$ or bounded measures are characterized by an H^p-condition in Knapp and Williamson [a] for $\lambda = \rho$ and in Michelson [b] for some more general eigenvalues. In Sjögren [a] another characterization using weak L^p-spaces is given.

Another related question is that of the irreducibility of $\mathcal{A}(G/K;\mathcal{M}_\lambda)$ as a representation space for G. This is settled in Helgason [g], where it is proved that $\mathcal{A}(G/K;\mathcal{M}_\lambda)$ is irreducible if and only if $e_\lambda e_{-\lambda} \neq 0$.

The global solvability of the inhomogeneous equation $Df = g$ $(f,g \in C^\infty(G/K)$, $D \in \mathbb{D}(G/K))$ is proved in Helgason [e]. For the Laplacian this is generalized to semisimple symmetric spaces in Chang [a] (and also in Kowata and Tanaka [a]).

Considering Theorem 5.4.3, it is natural to pose the problem, when $e_\lambda = 0$, to characterize the image of $\mathcal{P}_\lambda$ by some extra differential equations in addition to $\mathcal{M}_\lambda$. Some particular cases of this question have been answered in Johnson and Korányi [a], Berline and Vergne [a], and Johnson [c], [d], using generalizations of operators originally constructed by L. K. Hua [a] (see also Korányi and Malliavin [a] and Johnson [a], [b]). See also Lassalle [a], [b].

Analogs to Theorem 5.4.1 for spaces that are not Riemannian symmetric spaces of the noncompact type are given in Helgason [i] (the compact type), in Hashizume et al. [a], Helgason [f], Kowata and Okamoto [a], and Morimoto [b] (the Euclidean motion groups), in Helgason [l] (the Cartan motion groups), and in Hiraoka et al. [a], Oshima and Sekiguchi [a], Sekiguchi [a] and the announcement Oshima [b] (semisimple symmetric spaces).

See also Helgason [h], [k] and [n] for further references.

6. Boundary values on the full boundary

In the preceding chapter we have represented the joint eigenfunctions on G/K as Poisson integrals of their hyperfunction boundary values on K/M. When G/K has rank > 1 this is, however, only a small part of the boundary of G/K in $\tilde{X}$, and it is important to have analogous results for the other G-orbits in the boundary. In this chapter we therefore generalize the results of Chapter 5 to this situation.

6.1 Partial Poisson transformations

In this section we generalize the Poisson transformation taking into account the other boundary orbits.

For $F \subset \Delta$ and $\mu \in (\mathfrak{a}_F)^*_c$ let $\mathcal{B}(G/B_F;L_\mu)$ denote the space of hyperfunctions f on G satisfying

$$(6.1) \qquad f(g\,m\,a\,n) = a^{\mu-\rho} f(g)$$

for all $g \in G$, $m \in M_F \cap K$, $a \in A_F$ and $n \in N_F$. Then $\mathcal{B}(G/B_F;L_\mu)$ is canonically identified with the space of hyperfunction valued sections of the line bundle L_μ on G/B_F associated with the character σ^F_μ on B_F given by

$$\sigma^F_\mu(m\,a\,n) = a^{\rho-\mu} .$$

There is a natural action of the algebra $\mathbb{D}(M_F/M_F \cap K)$ of invariant differential operators on the symmetric space $M_F/M_F \cap K$ on $\mathcal{B}(G/B_F;L_\mu)$ coming from the action of $U(\mathfrak{m}_F)^{M_F \cap K}$ on $\mathcal{B}(G)$ from the right. This follows from the fact that M_F normalizes N_F and centralizes A_F.

For $\gamma \in \mathfrak{a}(F)^*_c$ and $\mu \in (\mathfrak{a}_F)^*_c$ we denote by $\mathcal{B}(G/B_F;L_\mu;\mathcal{M}_\gamma)$ the subspace of $\mathcal{B}(G/B_F;L_\mu)$ consisting of the eigenfunctions for all $D \in \mathbb{D}(M_F/M_F \cap K)$ with eigenvalues $\chi_\gamma(D)$. For $\lambda \in \mathfrak{a}^*_c$ we will for shortness use the symbol $\mathcal{B}(F;\lambda)$ for $\mathcal{B}(G/B_F;L_{\lambda|\mathfrak{a}_F};\mathcal{M}_{\lambda|\mathfrak{a}(F)})$. Notice that $\mathcal{B}(\emptyset;\lambda) = \mathcal{B}(G/P;L_\lambda)$ and $\mathcal{B}(\Delta;\lambda) = \mathcal{A}(G/K;\mathcal{M}_\lambda)$, and also that $\mathcal{B}(F;w\lambda) = \mathcal{B}(F;\lambda)$ for $w \in W_F$.

Let $E \subset F \subset \Delta$, $\mu \in (\mathfrak{a}_E)^*_c$ and $f \in \mathcal{B}(G/B_E;L_\mu)$.

Definition. The Poisson integral with respect to F of f is the hyperfunction $\mathcal{P}_E^F f$ on G given by

$$(6.2) \qquad \mathcal{P}_E^F f(g) = \int_{M_F \cap K} f(gk)dk \quad .$$

From Proposition 3.2.1(ix) it follows that

$$\mathcal{P}_E^F f \in \mathcal{B}(G/B_F;L_{\mu|\mathfrak{a}_F}) \quad .$$

The transformation

$$\mathcal{P}_E^F : \mathcal{B}(G/B_E;L_\mu) \longrightarrow \mathcal{B}(G/B_F;L_{\mu|\mathfrak{a}_F})$$

is called the (partial) Poisson transformation.

Notice that the spaces $\mathcal{B}(G/B_E;L_\mu)$ carry natural G-actions and that $\mathcal{P}_E^F$ is a G-map. Also it is obvious that if $D \subset E \subset F$ then

$$(6.3) \qquad \mathcal{P}_E^F \circ \mathcal{P}_D^E = \mathcal{P}_D^F \quad .$$

Let $\mathcal{P}_E = \mathcal{P}_E^\Delta$ and $\mathcal{P}^F = \mathcal{P}_\emptyset^F$. The Poisson transformation given by (5.6) is $\mathcal{P} = \mathcal{P}_\emptyset^\Delta$.

Proposition 6.1.1 Let $\lambda \in \mathfrak{a}_c^*$ and let $f \in \mathcal{B}(E;\lambda)$. Then

$$\mathcal{P}_E^F f \in \mathcal{B}(F;\lambda) \quad .$$

Proof: Consider the hyperfunction $\tilde{f}$ on $K \times M_F$ defined by

$$\tilde{f}(k,m) = f(km) \quad .$$

From (6.1) it follows that

$$\tilde{f}(k,m\,a\,n) = a^{\lambda-\rho}\,\tilde{f}(k,m)$$

for $k \in K$, $m \in M_F$, $a \in A_E \cap A(F)$, and $n \in N_E \cap N(F)$. Consider also the function $(\mathcal{P}_E^F f)^\sim$ defined by

$$(\mathcal{P}_E^F f)^{\sim}(k,m) = \mathcal{P}_E^F f(km)$$

for $k \in K$, $m \in M_F$. Then

$$(\mathcal{P}_E^F f)^{\sim}(k,m) = \int_{M_F \cap K} \tilde{f}(k,ml)dl$$

which shows that we may view $(\mathcal{P}_E^F f)^{\sim}$ as the "Poisson transform of $\tilde{f}$ in the second variable". Therefore it is not serious to facilitate notation by assuming $F = \Delta$.

We have

$$\mathcal{P}_E f(g) = \int_K f(gk)dk$$

and for $D = \Gamma(u) \in \mathbb{D}(G/K)$ we get as in Proposition 5.1.2 that

$$D\,\mathcal{P}_E f(gK) = \int_K (u_r f)(gk)dk \quad .$$

We now use the decomposition

$$U(\mathfrak{g}) = (\mathfrak{k}_E)_c\, U(\mathfrak{g}) \oplus U(\mathfrak{m}_E + \mathfrak{a}_E) \oplus U(\mathfrak{g})(\mathfrak{n}_E)_c$$

(cf. (3.16)) to project u to $U(\mathfrak{m}_E + \mathfrak{a}_E)^{M_E \cap K}$. Since f is an eigenfunction for $U(\mathfrak{m}_E + \mathfrak{a}_E)^{M_E \cap K}$ from the right it follows that

$$D\,\mathcal{P}_E f(gK) = \chi_\lambda(D)\mathcal{P}_E f(gK) \quad . \qquad \square$$

Since $\mathcal{P} = \mathcal{P}_F \circ \mathcal{P}^F$ the following theorem is an immediate consequence of Corollary 5.4.4:

<u>Theorem 6.1.2</u> <u>Let</u> $u \in \mathcal{A}(G/K)$ <u>be a joint eigenfunction for</u> $\mathbb{D}(G/K)$ <u>and let</u> $F \subset \Delta$. <u>There exists</u> $\lambda \in \mathfrak{a}_c^*$ <u>and</u> $f \in \mathcal{B}(F;\lambda)$ <u>such that</u> $u = \mathcal{P}_F f$.

Later on (Theorem 6.4.1) we shall see that if λ is chosen such that $\mathrm{Re}\langle\lambda,\alpha\rangle \geq 0$ for all $\alpha \in \Sigma^+ \setminus \langle F\rangle$, then f is unique.

We will now use the transformation $\mathcal{P}^F = \mathcal{P}_\emptyset^F$ to obtain an analog of Theorem 5.1.4. We need the following lemma, which in fact enters into the proof of Proposition 5.1.3. For $s \in W$ let $\Sigma^+(s) = \{\alpha \in \Sigma^+ \mid s\alpha \in -\Sigma^+\}$, $\bar{\mathfrak{n}}_s = \sum_{\alpha \in \Sigma^+(s)} \mathfrak{g}^{-\alpha}$, and $\overline{N}_s = \exp \overline{\mathfrak{n}}_s = \overline{N} \cap s^{-1}Ns$.

Lemma 6.1.3 Let $\lambda \in \mathfrak{a}_c^*$ and $s \in W$. If $\mathrm{Re}\langle\lambda,\alpha\rangle > 0$ for all $\alpha \in \Sigma^+(s)$ then the integral

$$c_\lambda(s) = \int_{\overline{N}_s} \exp\langle -\lambda-\rho , H(\overline{n})\rangle \, d\overline{n}$$

converges absolutely. The value $c_\lambda(s)$ of the integral is given by the expression (5.7), the product taken only over $\Sigma_o^+ \cap \Sigma^+(s)$.

Proof: (Gindikin and Karpelevič [a]) See Helgason [n] Chapter 4, Theorem 6.13. □

Lemma 6.1.4 Let $\lambda \in \mathfrak{a}_c^*$. If $\mathrm{Re}\langle\lambda,\alpha\rangle > 0$ for all $\alpha \in \Sigma^+ \backslash \langle F\rangle$ then the integral

$$(6.4) \qquad c_\lambda^F = \int_{\overline{N}_F} \exp\langle -\lambda-\rho , H(\overline{n})\rangle \, d\overline{n}$$

converges absolutely. If $\mathrm{Re}\langle\lambda,\alpha\rangle > 0$ for all $\alpha \in \Sigma^+ \cap \langle F\rangle$ then the integral

$$(6.5) \qquad c_\lambda(F) = \int_{\overline{N}(F)} \exp\langle -\lambda-\rho , H(\overline{n})\rangle \, d\overline{n}$$

converges absolutely. The value c_λ^F , respectively $c_\lambda(F)$, is given by the expression (5.7), the product taken only over those roots $\alpha \in \Sigma_o^+$ for which $\alpha \notin \langle F\rangle$, respectively $\alpha \in \langle F\rangle$. In particular $c_\lambda = c_\lambda^F c_\lambda(F)$.

Proof: Let s_o , resp. s_o' , be the unique element of W , resp. W_F , such that $s_o\Sigma^+ = -\Sigma^+$, resp. $s_o'(\Sigma^+ \cap \langle F\rangle) = -\Sigma^+ \cap \langle F\rangle$. Then $\Sigma^+(s_o') = \Sigma^+ \cap \langle F\rangle$ and $\Sigma^+(s_o s_o') = \Sigma^+ \backslash \langle F\rangle$. Therefore this lemma follows from the preceding one. □

Henceforth c_λ^F and $c_\lambda(F)$ denote the meromorphic functions in $\lambda \in \mathfrak{a}_c^*$ given by this lemma.

In the following theorem and corollary the notation $a \xrightarrow[F]{} \infty$ means that $a \in A_F$ and $a^\alpha \longrightarrow \infty$ for all $\alpha \in \Delta \backslash F$.

Theorem 6.1.5 Let $\lambda \in \mathfrak{a}_c^*$ be such that $\mathrm{Re}\langle\lambda,\alpha\rangle > 0$ for all $\alpha \in \Sigma^+ \setminus \langle F\rangle$ and let $f \in \mathcal{B}(G/P;L_\lambda)$ be continuous. Then

$$\mathcal{P}^F f(x) = (c_\lambda^F)^{-1} \lim_{a \xrightarrow[F]{} \infty} a^{\rho-\lambda} \mathcal{P} f(xa) \tag{6.6}$$

for each $x \in G$. The convergence is uniform in x on compact sets.

Proof: As in the proof of Theorem 5.1.4, we have

$$a^{\rho-\lambda} \mathcal{P} f(xa) = \int_{\overline{N}} f(xa\bar{n}a^{-1}) \exp\langle -\lambda-\rho, H(\bar{n})\rangle d\bar{n} \quad . \tag{6.7}$$

Let $\bar{n} = \bar{n}_1\bar{n}'$ where $\bar{n}_1 \in \overline{N}_F$ and $\bar{n}' \in \overline{N}(F)$, then we see that (6.7) equals

$$\int_{\overline{N}(F)} \int_{\overline{N}_F} f(xa\bar{n}_1 a^{-1} \bar{n}')\exp\langle -\lambda-\rho , H(\bar{n}_1\bar{n}')\rangle d\bar{n}_1 d\bar{n}'$$

$$= \int_{\overline{N}(F)} \int_{\overline{N}_F} f(xa\bar{n}_1 a^{-1}\kappa(\bar{n}'))\exp\langle -2\rho, H(\bar{n}')\rangle \exp\langle -\lambda-\rho, H(\bar{n}_1\kappa(\bar{n}'))\rangle d\bar{n}_1 d\bar{n}'$$

$$= \int_{M_F \cap K} \int_{\overline{N}_F} f(xa\bar{n}_1 a^{-1}k)\exp\langle -\lambda-\rho , H(\bar{n}_1 k)\rangle d\bar{n}_1 dk$$

$$= \int_{M_F \cap K} \int_{\overline{N}_F} f(xka\bar{n}_1 a^{-1})\exp\langle -\lambda-\rho , H(\bar{n}_1)\rangle d\bar{n}_1 dk \quad .$$

From the definition of $a \xrightarrow[F]{} \infty$ it follows that $a\bar{n}_1 a^{-1} \longrightarrow e$ as $a \xrightarrow[F]{} \infty$, and since

$$|f(xka\bar{n}_1 a^{-1})| \le \exp\langle \mathrm{Re}\,\lambda-\rho , H(a\bar{n}_1 a^{-1})\rangle \sup_{y \in K} |f(xy)|$$

the theorem follows by dominated convergence once we prove that

$$\varphi_a(\bar{n}_1) = \exp\langle \mathrm{Re}\,\lambda-\rho , H(a\bar{n}_1 a^{-1})\rangle \exp\langle -\mathrm{Re}\,\lambda-\rho , H(\bar{n}_1)\rangle$$

is uniformly in $L^1(\overline{N}_F)$ as $a \xrightarrow[F]{} \infty$.

If $\mathrm{Re}\,\lambda \in \mathfrak{a}_+^*$ we can repeat the argument from the proof of Theorem 5.1.4, but under our slightly more general assumption on λ we need to generalize (5.9). This is done in the following lemma. From this the theorem then follows as in Theorem 5.1.4. □

Let $\mathfrak{a}_F^+ = \{H \in \mathfrak{a}_F \mid \alpha(H) > 0 \text{ for all } \alpha \in \Delta\setminus F\}$ and let $A_F^+ = \exp \mathfrak{a}_F^+$.

Lemma 6.1.6 Let $\nu \in \mathfrak{a}^*$ be such that $\langle\nu,\alpha\rangle \ge 0$ for all $\alpha \in \Sigma^+ \setminus \langle F\rangle$, and let $a \in A_F^+$, $\bar{n} \in \overline{N}$, $\bar{n}_1 \in \overline{N}_F$ and $k \in K$. Then the following holds:

(i) $\langle \nu, H(\bar{n}_1) \rangle \geq 0$,

(ii) $\langle \nu, H(\bar{n}) - H(a\bar{n}a^{-1}) \rangle \geq 0$,

(iii) $\langle \nu, H(a) - H(ak) \rangle \geq 0$.

Proof: (i) Follows from (ii) by taking $\bar{n} = \bar{n}_1$ and letting $a \underset{F}{\longrightarrow} \infty$.

(ii) Follows from (iii) since

$$H(\bar{n}) - H(a\bar{n}a^{-1}) = H(a) - H(a\kappa(\bar{n})) \quad .$$

(iii) By Kostant's convexity theorem (Kostant [a] Theorem 4.1 or Helgason [n] Chapter 4, Section 10) it suffices to prove that

$$\langle \nu, H - wH \rangle \geq 0$$

for all $w \in W$ and $H \in \mathfrak{a}_F^+$. From the assumption on ν it follows that $s\nu \in \mathfrak{a}_+^*$ for some $s \in W_F$, and then

$$\langle s\nu, H - wH \rangle \geq 0$$

for all $w \in W$ (see p. 91). Since $sH = H$ the lemma follows. □

Remark. The proof of Theorem 6.1.5 actually shows that (6.6) holds whenever $f \in \mathcal{B}(G/P;L_\lambda)$ such that $f|_K \in L^\infty(K)$ and f is continuous at each point xk , $k \in M_F \cap K$.

From Theorem 6.1.5 one gets the following corollary, similar to Corollary 5.1.5. The proof is also similar (using Lemma 6.2.4 below), so we omit it.

Corollary 6.1.7 Let $\lambda \in \mathfrak{a}_c^*$ be such that $\mathrm{Re}\langle \lambda, \alpha \rangle > 0$ for all $\alpha \in \Sigma^+ \setminus \langle F \rangle$, and let $f \in \mathcal{B}(G/P;L_\lambda)$. Then for each $m \in M_F$ and each analytic function $\varphi \in \mathcal{A}(K)$

$$\int_K \mathcal{P}^F f(km)\varphi(k)\,dk = (c_\lambda^F)^{-1} \lim_{a \underset{F}{\longrightarrow} \infty} a^{\rho-\lambda} \int_K \mathcal{P} f(kma)\varphi(k)\,dk \ .$$

6.2 Partial spherical functions and Poisson kernels

Let $F \subset \Delta$ and $\lambda \in \mathfrak{a}_c^*$. Define $\phi_\lambda^F \in \mathcal{B}(F;\lambda)$ as the Poisson integral $\mathcal{P}^F$ of the constant function 1 on K/M, i.e.,

(6.8) $$\phi_\lambda^F(g) = \int_{M_F \cap K} \exp\langle \lambda-\rho, H(gk)\rangle dk .$$

We call ϕ_λ^F the partial spherical function with parameter λ. It can be uniquely characterized as follows:

Lemma 6.2.1 The partial spherical function ϕ_λ^F is up to constants the unique element in $\mathcal{B}(F;\lambda)$ which is K-invariant.

Proof: Any K-invariant element in $\mathcal{B}(F;\lambda)$ is uniquely determined by its restriction to M_F, where it is a spherical function and hence unique. □

From Lemma 5.3.1 applied to $\phi_\lambda^F|_{M_F}$ it follows that $\phi_\mu^F = \phi_\lambda^F$ if and only if $\mu = w\lambda$ for some $w \in W_F$.

When λ satisfies a certain regularity condition we can give another characterization of ϕ_λ^F. From (6.8) it follows that the function $x \longrightarrow \phi_{-\lambda}^F(x^{-1}) = \int_{M_F \cap K} P_\lambda(x,k)dk$ belongs to $\mathcal{A}(G/K;\mathcal{M}_\lambda)$.

Proposition 6.2.2 Assume $(w\lambda-\lambda)|_{\mathfrak{a}_F} \neq 0$ for all $w \in W \setminus W_F$. Then the space of $(M_F \cap K)N_F$-invariant elements in $\mathcal{A}(G/K;\mathcal{M}_\lambda)$ is spanned by the functions $\phi_{-w\lambda}^F(x^{-1})$ for $w \in W$.

Proof: Let I_λ denote the space of $(M_F \cap K)N_F$-invariant elements in $\mathcal{A}(G/K;\mathcal{M}_\lambda)$, and let $r = |W|/|W_F|$. We will prove that $\dim I_\lambda \leq r$ for all $\lambda \in \mathfrak{a}_c^*$. Since the given assumption on λ implies that the functions $\phi_{-w\lambda}^F$ for $w \in W_F \setminus W$ are linearly independent, this will prove the proposition.

Recall from Section 3.3 the map $\delta_F : U(\mathfrak{g})^K \longrightarrow U(\mathfrak{m}_F + \mathfrak{a}_F)^{M_F \cap K}$ given by

$$u - \delta_F(u) \in (\mathfrak{n}_F)_c U(\mathfrak{g}) + U(\mathfrak{g})(\mathfrak{q}_F)_c$$

where $\mathfrak{q}_F$ denotes the orthocomplement of $\mathfrak{m}_F \cap \mathfrak{k}$ in $\mathfrak{k}$ (we have exchanged $\bar{\mathfrak{n}}_F$ with $\mathfrak{n}_F$ from (3.16)). From (3.19) we have

(6.9) $U(\mathfrak{m}_F + \mathfrak{a}_F)^{M_F \cap K} \subset \Sigma_{i=1}^r \nu_i \delta_F(U(\mathfrak{g})^K) + U(\mathfrak{m}_F + \mathfrak{a}_F)(\mathfrak{m}_F \cap \mathfrak{h})_c$.

For $f \in I_\lambda$, $m \in M_F A_F$, and $u \in U(\mathfrak{g})^K$ we have

$$(\delta_F(u)_r f)(m) = (u_r f)(m) = \chi_\lambda(u) f(m)$$

since m normalizes N_F. Hence the restriction $\bar{f} = f|_{M_F A_F}$ is an eigenfunction for $\delta_F(U(\mathfrak{g})^K)$.

We now claim that if $\nu_j \bar{f}(e) = 0$ for all $j = 1,\ldots,r$, then $f = 0$. In fact it follows from (6.9) that then $v\bar{f}(e) = 0$ for all $v \in U(\mathfrak{m}_F + \mathfrak{a}_F)^{M_F \cap K}$, and since $\bar{f}$ is biinvariant for $M_F \cap K$ this implies that $\bar{f}$ and hence f vanish. Hence $f \in I_\lambda$ is uniquely determined by the values $\nu_1 \bar{f}(e),\ldots,\nu_r \bar{f}(e)$ and thus $\dim I_\lambda \leq r$. □

Corollary 6.2.3 Assume $(w\lambda - \lambda)|_{\mathfrak{a}_F} \neq 0$ for all $w \in W \setminus W_F$. Let $\mu \in (\mathfrak{a}_F)_c^*$ and $f \in \mathcal{A}(G/K; \mathcal{M}_\lambda)$ and assume

$$f(bx) = \exp\langle \mu + \rho, H(b) \rangle f(x)$$

for all $b \in B_F$, $x \in G/K$. If $f \neq 0$ there exists $w \in W$ and $c \in \mathbb{C}$ such that $\mu = (w\lambda)|_{\mathfrak{a}_F}$ and $f(x) = c\, \phi^F_{-w\lambda}(x^{-1})$.

Thus ϕ^F_λ is up to scalars uniquely determined in $\mathcal{B}(G/B_F; L_\lambda|_{\mathfrak{a}_F})$ by the property that $x \longrightarrow \phi^F_\lambda(x^{-1})$ belongs to $\mathcal{A}(G/K; \mathcal{M}_{-\lambda})$.

We will now construct another important element of $\mathcal{B}(F;\lambda)$. Let $\delta \in \mathcal{B}(K/M)$ be the Dirac measure supported at the origin, and define $S^F_\lambda \in \mathcal{B}(F,\lambda)$ by

$$S^F_\lambda = \mathcal{P}^F \delta \quad .$$

Using the canonical identification of $\mathcal{B}(K/M)$ with $\mathcal{B}(G/P; L_\lambda)$ it is easily seen that S^F_λ is the distribution on G given by

$$S^F_\lambda(\varphi) = \int_{P_F} \varphi(p_F) \exp\langle -\lambda-\rho, H(p_F^{-1}) \rangle dp_F \tag{6.10}$$

for $\varphi \in C_c^\infty(G)$, where dp_F is left invariant Haar measure on P_F:

$$\int_{P_F} h(p_F) dp_F = \int_{M_F} \int_{A_F} \int_{N_F} h(m\,a\,n)\,dm\,da\,dn$$

suitably normalized.

Let $P_\lambda^F(k,x) = S_\lambda^F(k^{-1}x)$ for $k \in K$, $x \in G$.

Lemma 6.2.4 For each $\lambda \in \mathfrak{a}_c^*$ and $f \in \mathcal{B}(G/P;L_\lambda)$ we have

$$\mathcal{P}^F f(x) = \int_K f(k) P_\lambda^F(k,x)\,dk$$

for $x \in G$ (as an equation of hyperfunctions).

Proof: For $F = \emptyset$ the equation says that

(6.11) $$f(x) = \int_K f(k)\,\delta(k^{-1}x)\,dk$$

which is obvious. If we apply $\mathcal{P}^F$ to both sides of (6.11) we get the lemma. □

Because of this lemma we call the distribution $P_\lambda(k,x)$ the partial Poisson kernel.

Notice that

$$S_\lambda^\Delta(g) = \phi_{-\lambda}^\emptyset(g) = \exp\langle -\lambda-\rho, H(g)\rangle .$$

There is another remarkable relation between the functions ϕ_λ^F and S_λ^F :

Lemma 6.2.5 $\phi_\lambda^F(g) = \int_K S_\lambda^F(kg)\,dk$ for $g \in G$ and $\lambda \in \mathfrak{a}_c^*$.

Proof: Follows immediately from Lemma 6.2.4 since $\phi_\lambda^F = \mathcal{P}^F 1$. □

From (6.10) it follows that

(6.12) $$S_\lambda^F(pg) = \exp\langle \lambda+\rho, H(p)\rangle S_\lambda^F(g)$$

for $p \in P$ and $g \in G$, i.e., $S_\lambda^F(x^{-1})$ belongs to $\mathcal{B}(G/P;L_{-\lambda})$. The preceding lemma then says that $\phi_\lambda^F(x^{-1})$ is the Poisson transform of $S_\lambda^F(x^{-1})$.

We will now show that (6.12) determines S_λ^F uniquely provided λ satisfies some regularity conditions.

Proposition 6.2.6 Let $\lambda \in \mathfrak{a}_c^*$ and assume that $e_{-\lambda} \neq 0$ (cf. (5.17)) and that $(w\lambda-\lambda)|_{\mathfrak{a}_F} \neq 0$ for all $w \in W\setminus W_F$. Let $\mu \in (\mathfrak{a}_F)_c^*$ and $f \in \mathcal{B}(G/B_F;L_\mu)$, and assume that f satisfies

$$f(pg) = \exp\langle\lambda+\rho\,,\,H(p)\rangle f(g)$$

<u>for all</u> $p \in P\,,\ \ g \in G$.

(i) <u>If</u> $\mu \neq (w\lambda)|_{\mathfrak{a}_F}$ <u>for all</u> $w \in W$ <u>then</u> $f = 0$.

(ii) <u>If</u> $\mu = \lambda|_{\mathfrak{a}_F}$ <u>then</u> $f = cS_\lambda^F$ <u>for some constant</u> $c \in \mathbb{C}$.

<u>Proof</u>: Let $f^\vee(g) = f(g^{-1})$, then $f^\vee \in \mathcal{B}(G/P;L_{-\lambda})$ and

$$f^\vee(bg) = \exp\langle -\mu+\rho\,,\,H(b)\rangle f(g)$$

for $b \in B_F$, $g \in G$. From Corollary 6.2.3 it follows that $\mathcal{P}_{-\lambda}f^\vee = 0$ in case of (i) and $\mathcal{P}_{-\lambda}f^\vee = c\phi_\lambda^{F\vee}$ in case of (ii) . Since $\mathcal{P}_{-\lambda}$ is injective by Theorem 5.4.3, the proposition follows. □

6.3 <u>Boundary values and asymptotics</u>

Let $F \subset \Delta$ and $\lambda \in \mathfrak{a}_c^*$, and let $u \in \mathcal{A}(G/K;\mathcal{M}_\lambda)$. In this section we show, under a certain regularity condition on λ , that a hyperfunction $f \in \mathcal{B}(F;\lambda)$ whose Poisson integral $\mathcal{P}_F f$ is u (cf. Theorem 6.1.2) can be viewed as a boundary value of u .

The assumption on λ is

<u>Assumption $(A)^F$</u>

$$\langle w\lambda-\lambda\,,\,H_j\rangle \notin \mathbb{Z} \quad \underline{\text{for all}}\ \ \alpha_j \notin F \ \ \underline{\text{and}}\ \ w \in W\setminus W(H_j)\,.$$

(Recall that $W(H_j) = \{w \in W \mid wH_j = H_j\}$) . Notice that Assumption $(A)^\emptyset$ is considerably stronger than Assumption (A), because it requires $\langle w\lambda-\lambda\,,\,H_j\rangle \notin \mathbb{Z}$ for <u>all</u> j such that $wH_j \neq H_j$.

<u>Lemma 6.3.1</u> Let $\lambda \in \mathfrak{a}_c^*$ <u>and assume</u> $(A)^F$. <u>Then there exists</u> $w \in W_F$ <u>such that</u> $e_{w\lambda} \neq 0$.

<u>Proof</u>: From the first part of the proof of Lemma 5.2.3 it follows that $2\langle\lambda,\alpha\rangle/\langle\alpha,\alpha\rangle$ is not an integer for all $\alpha \in \Sigma\setminus\langle F\rangle$. On the other hand we can choose $w \in W_F$ such that $2\langle w\lambda,\alpha\rangle/\langle\alpha,\alpha\rangle$ is not a negative integer for all $\alpha \in \Sigma^+ \cap \langle F\rangle$. Hence the lemma. □

Since replacing λ by $w\lambda$ for $w \in W_F$ has no influence on the set $\mathcal{B}(F;\lambda)$ or the map $\mathcal{P}_F$, we may assume $e_\lambda \neq 0$ in addition to $(A)^F$.

Fix $g \in G$ and let φ_g be the corresponding system of local coordinates on $\widetilde{X}$. By Remark 4.3.3 the system of operators $\{Q_j \mid \alpha_j \notin F\}$ has regular singularities in the weak sense along the walls N_j $(\alpha_j \notin F)$ with the edge

$$\{\pi(g\bar{n}, t) \mid \bar{n} \in \bar{N}, \ t \in \mathbb{R}_+^F\} \ .$$

(Recall that $\mathbb{R}_+^F = \{t \in \mathbb{R}^n \mid t_j > 0 \ \text{if} \ \alpha_j \in F, \ t_j = 0 \ \text{otherwise}\}$). The characteristic exponents are given by (4.24)

$$s_{\bar{\sigma}} = (\langle \rho - \sigma_j\lambda, H_j \rangle)_{\alpha_j \notin F} \ \in \ \mathbb{C}^p$$

where p is the number of elements in $\Delta\setminus F$ and σ is a p-tuple $\sigma = (\sigma_j)_{\alpha_j \notin F}$ of elements σ_j in W. Finally $\bar{\sigma}$ denotes the right coset

$$\bar{\sigma} = (W(H_j)\sigma_j)_{\alpha_j \notin F} \ .$$

We see that Assumption $(A)^F$ is equivalent to

$$s_{\bar{\sigma}} - s_{\bar{\sigma}'} \notin \mathbb{Z}^p$$

for all right cosets $\bar{\sigma} \neq \bar{\sigma}'$.

By the theory of Chapter 2 u has boundary values on the edge. Though the operators Q_j themselves (and not just their local expressions) vary from one system of local coordinates to another (i.e., they depend on g), the characteristic exponents do not vary. By Theorems 2.5.4 and 2.5.8 the boundary values are defined on

$$G/B_F \cong \pi(G \times \mathbb{R}_+^F)$$

as sections of the line bundle

$$\bigotimes_{\alpha_j \notin F} (T^*_{N_j}\widetilde{X})^{\otimes\langle\rho-\sigma_j\lambda, H_j\rangle}$$

for each right coset $\bar{\sigma}$ as above.

Using Lemma 5.2.1 it follows by analogy with Corollary 5.2.2 that the boundary value map results in a G-map

$$\beta_{F,\bar{\sigma},\lambda} : \mathcal{A}(G/K;\mathcal{M}_\lambda) \longrightarrow \mathcal{B}(G/B_F;L_{\mu(\bar{\sigma},\lambda)})$$

where $\mu(\bar{\sigma},\lambda) \in (\mathfrak{a}_F)_c^*$ is given by

$$<\mu(\bar{\sigma},\lambda)\,,\,H_j> \;=\; <\sigma_j\lambda\,,\,H_j>$$

for $\alpha_j \notin F$.

<u>Lemma 6.3.2</u> <u>Assume</u> $(A)^F$ <u>and let</u> $\bar{\sigma} = (W(H_j)\sigma_j)_{\alpha_j \notin F}$ <u>where</u> $\sigma_j \in W$ for $\alpha_j \notin F$.

(i) <u>The boundary value map</u> $\beta_{F,\bar{\sigma},\lambda}$ <u>is identically zero unless there exists</u> $w \in W$ <u>such that</u> $\sigma_j \in W(H_j)w$ <u>for all</u> $\alpha_j \notin F$.

(ii) <u>If</u> $\sigma_j \in W(H_j)w$ <u>for all</u> $\alpha_j \notin F$ <u>for some</u> $w \in W$ <u>then there exists a complex number</u> $c = c(F,w\lambda)$ <u>such that</u>

$$\beta_{F,\bar{\sigma},\lambda}\,\mathcal{P}_{w\lambda}f = c\,\mathcal{P}^F f \tag{6.13}$$

<u>for all</u> $f \in \mathcal{B}(G/P;L_{w\lambda})$.

<u>Proof</u>: Let $s \in W$ and $\varphi \in \mathcal{B}(K/M)$. Let

$$U(k,x) = \varphi(k)\exp<-s\lambda-\rho\,,\,H(x^{-1}k)> \;=\; \varphi(k)S^{\Delta}_{s\lambda}(k^{-1}x)$$

for $k \in K$, $x \in G$. From Lemma 5.4.2 we get

$$\beta_{F,\bar{\sigma},\lambda}\,\mathcal{P}_{s\lambda}\,\varphi(x) = \int_K \varphi(k)\;\beta_{F,\bar{\sigma},\lambda}\,S^{\Delta}_{s\lambda}(k^{-1}x)dk\,. \tag{6.14}$$

We notice that since $\beta_{F,\bar{\sigma},\lambda}$ is a G-map we have

$$\beta_{F,\bar{\sigma},\lambda}S^{\Delta}_{s\lambda}(pg) = \exp<w\lambda+\rho\,,\,H(p)>\beta_{F,\bar{\sigma},\lambda}S^{\Delta}_{s\lambda}(g)$$

for all $p \in P$, $g \in G$.

(i) Assume $\beta_{F,\bar{\sigma},\lambda} \neq 0$. By Lemma 6.3.1 we may assume $e_\lambda \neq 0$, and hence $\mathcal{P}_\lambda$ is surjective by Theorem 5.4.3. It follows from (6.14) with $s=e$ that

$$\beta_{F,\bar{\sigma},\lambda}\,S^{\Delta}_{\lambda} \neq 0\,. \tag{6.15}$$

Since $\beta_{F,\bar{\sigma},\lambda}S^{\Delta}_{\lambda}$ is analytic in λ we have (6.15) for some λ satisfying $e_{-\lambda} \neq 0$, and then it follows from Proposition 6.2.6(i) that $\mu(\bar{\sigma},\lambda) = (w\lambda)|_{\mathfrak{a}_F}$ for some $w \in W$, which proves (i).

(ii) Assume $\sigma_j \in W(H_j)w$ for all $\alpha_j \notin F$. Then $\mu(\bar{\sigma},\lambda) = (w\lambda)|_{\mathfrak{a}_F}$ and hence

$$\beta_{F,\bar{\sigma},\lambda}\, S^{\Delta}_{w\lambda} = c\, S^{F}_{w\lambda}$$

for some $c \in \mathbb{C}$ by Proposition 6.2.6(ii). Then (6.14) and Lemma 6.2.4 proves (6.13). It easily follows from (6.13) that c remains the same if we exchange w and λ by wv^{-1} and $v\lambda$, respectively, for any $v \in W$, i.e., $c = c(F, w\lambda)$. □

In particular it follows from (6.13) that $\beta_{F,\bar{\sigma},\lambda}$ maps $\mathcal{A}(G/K;\mathcal{M}_\lambda)$ to $\mathcal{B}(F;w\lambda)$, because $\mathcal{P}_{sw\lambda}$ is surjective by Theorem 5.4.3, when $s \in W_F$ is chosen such that $e_{sw\lambda} \neq 0$ (cf. Lemma 6.3.1).

We thus essentially have r maps $(r = |W|/|W_F|)$:

$$\beta_{F,w\lambda} : \mathcal{A}(G/K;\mathcal{M}_\lambda) \longrightarrow \mathcal{B}(F;w\lambda) \tag{6.16}$$

for w in a set of representatives for $W_F\backslash W$.

<u>Theorem 6.3.3</u> <u>Let</u> $\lambda \in \mathfrak{a}^*_c$ <u>and</u> $F \subset \Delta$. <u>If Assumption</u> $(A)^F$ <u>holds then</u>

$$\mathcal{P}_F : \mathcal{B}(F;\lambda) \longrightarrow \mathcal{A}(G/K;\mathcal{M}_\lambda)$$

<u>is bijective and its inverse is</u> $(c^F_\lambda)^{-1}\beta_{F,\lambda}$, <u>where</u> c^F_λ <u>is given by (6.4)</u>.

<u>Proof</u>: Assuming $e_\lambda \neq 0$ we have that $\mathcal{P} = \mathcal{P}_\lambda$ is bijective. From (6.13) we get

$$\beta_{F,\lambda} = c(F,\lambda)\mathcal{P}^F \circ \mathcal{P}^{-1}$$

and hence since $\mathcal{P} = \mathcal{P}_F \circ \mathcal{P}^F$

$$\beta_{F,\lambda} \circ \mathcal{P}_F = \mathcal{P}_F \circ \beta_{F,\lambda} = c(F,\lambda) \quad .$$

It therefore only remains to identify $c(F,\lambda)$ with c^F_λ. This is done in the proof of the following theorem. □

Fix $F \subset \Delta$ and let $\mathfrak{a}^{*\prime\prime}_c$ denote the set of all $\lambda \in \mathfrak{a}^*_c$ satisfying $(A)^F$.

<u>Theorem 6.3.4</u> <u>There exists, for each</u> $\lambda \in \mathfrak{a}^{*\prime\prime}_c$, <u>an open set</u> $\Omega \subset \mathbb{R}^n$ <u>containing</u> $\mathbb{R}^F_+$, <u>and for each right coset</u> $\bar{w} = W_F w \in W_F\backslash W$ <u>a function</u> $h^F_{\bar{w}}(\lambda, t)$ <u>of</u> $t \in \Omega$ <u>such that the following statements hold.</u>

(i) <u>For</u> $t \in \mathbb{R}_+^n \cap \Omega$ <u>we have</u>

$$\phi_\lambda(a_t K) = \Sigma_{\bar{w} \in W_F \backslash W}\, h_{\bar{w}}^F(\lambda, t) \prod_{\alpha_j \notin F} t_j^{<\rho - w\lambda, H_j>} .$$

(ii) $h_{\bar{w}}^F(\lambda, t)$ <u>is real analytic in</u> $t \in \Omega$ <u>and holomorphic in</u> $\lambda \in \mathfrak{a}_c^{*\prime\prime}$.

(iii) <u>For</u> $t \in \mathbb{R}_+^F$ <u>we have</u> $h_{\bar{w}}^F(\lambda, t) = c_{w\lambda}^F\, \phi_{w\lambda}^F(a_t)$ <u>where</u> $\phi_{w\lambda}^F$ <u>is given by (6.8)</u>.

(Recall that $a_t \in A$ for $t \in \mathbb{R}_+^n$ is defined by

$$a_t = \exp(\Sigma_{t_j \neq 0} - (\log t_j)H_j)$$

cf. (4.1).)

<u>Proof</u>: From (6.13) we have $\beta_{F,w\lambda}\, \phi_\lambda = c(F, w\lambda)\phi_{w\lambda}^F$. Thus ϕ_λ has real analytic boundary values and Theorem 2.5.6 can be applied. This gives the theorem (cf. the proof of Theorem 5.3.2), except for the identification of $c(F, \lambda)$ with c_λ^F. For this we assume $\mathrm{Re} < \lambda, \alpha > \, > 0$ for all $\alpha \in \Sigma^+$ and get from Theorem 6.1.5 that

$$a^{\rho - \lambda}\, \phi_\lambda(xa) \longrightarrow c_\lambda^F\, \phi_\lambda^F(x) \tag{6.17}$$

as $a \xrightarrow[F]{} \infty$. Then $c_\lambda^F = c(F, \lambda)$ follows in analogy with Proposition 5.3.3. □

<u>Remark.</u> Under the hypothesis that λ satisfies $(A)^{\emptyset}$, the results proved in this section do not depend upon Theorem 5.4.3, of which proof was omitted.

6.4 <u>The bijectivity of the partial Poisson transformations</u>

Let $E \subset F \subset \Delta$ and let $e_\lambda^E(F)^{-1}$ denote the following meromorphic function on $\mathfrak{a}_c^*$:

$$e_\lambda^E(F)^{-1} = \prod_{\alpha \in \Sigma_0^+ \cap <F> \backslash <E>} \Gamma\left(\frac{1}{2}\left(\frac{1}{2}m(\alpha) + 1 + \frac{<\lambda,\alpha>}{<\alpha,\alpha>}\right)\right)\Gamma\left(\frac{1}{2}\left(\frac{1}{2}m(\alpha) + m(2\alpha) + \frac{<\lambda,\alpha>}{<\alpha,\alpha>}\right)\right)$$

then e_λ as given by (5.17) equals $e_\lambda^\emptyset(\Delta)$, and we have $e_\lambda^D(F) = e_\lambda^D(E)e_\lambda^E(F)$ for $D \subset E \subset F$.

Theorem 6.4.1 Let $\lambda \in \mathfrak{a}_c^*$ and $\mathcal{P}_E^F : \mathcal{B}(E;\lambda) \longrightarrow \mathcal{B}(F;\lambda)$ the partial Poisson transformation. The following statements are equivalent.

(i) $e_\lambda^E(F) \neq 0$.

(ii) $\mathcal{P}_E^F$ is injective.

(iii) $\mathcal{P}_E^F$ is surjective.

Proof: The theorem is first reduced to the case where $E = \emptyset$. Replacing λ by $w\lambda$ for a suitable $\lambda \in W_E$ we may assume that $e_\lambda(E) = e_\lambda^\emptyset(E) \neq 0$. Assuming the theorem proved for $\mathcal{P}_\emptyset^D$ for all $D \subset \Delta$ we apply this to $D = E$ and get that $\mathcal{P}_\emptyset^E$ is bijective. Therefore (i), (ii) and (iii) are equivalent to the similar statements for $\mathcal{P}_\emptyset^F$. Thus we may assume $E = \emptyset$.

For $f \in \mathcal{B}(K/M)$ and $K \in K$, $m \in M_F$ we have

$$\mathcal{P}^F f(km) = \int_{M_F \cap K} f(kl) \exp\langle -\lambda-\rho , H(m^{-1}l)\rangle dl$$

by Lemma 5.1.1. We see that $\mathcal{P}^F f|_{KM_F}$ only depends on $\lambda|_{\mathfrak{a}(F)}$. Therefore, altering $\lambda|_{\mathfrak{a}_F}$ if necessary, we may assume that Assumption $(A)^F$ holds. Then $\mathcal{P}_F$ is bijective by Theorem 6.3.3, and hence (ii) and (iii) are equivalent to the similar statements for $\mathcal{P}_\lambda$. Also, by Lemma 6.3.1, $(A)^F$ implies that $c_\lambda^F \neq 0$ because $e_{w\lambda} = e_{w\lambda}^\emptyset(F)e_\lambda^F(\Delta)$ for $w \in W_F$. Therefore (i) is equivalent to $e_\lambda \neq 0$. Then the theorem follows from Theorem 5.4.3. □

If λ satisfies that $\langle w\lambda-\lambda, H_i\rangle \notin \mathbb{Z}$ for each $\alpha_j \in F \setminus E$ and $w \in W_F \setminus W_E$ we can in fact construct the inverse of $\mathcal{P}_E^F$ as a boundary value map. For $f \in \mathcal{B}(F;\lambda)$ we define $\tilde{f} \in \mathcal{B}(K \times M_F)$ by $\tilde{f}(k,m) = f(km)$. Then $\tilde{f}$ belongs to $\mathcal{A}(M_F/M_F \cap K; \mathcal{M}_{\lambda|_{\mathfrak{a}(F)}})$ in the second variable, and we can define the boundary value $\beta_{E,\lambda}^F \tilde{f}$ in

$$\mathcal{B}(K\times(M_F/M_F\cap B_E);L_{\lambda|_{\mathfrak{a}_E}}) = \left\{ f \in \mathcal{B}(K\times M_F) \;\middle|\; \begin{array}{l} f(k,x\,man) = a^{\lambda-\rho}f(k,x) \quad \text{for} \\ k\in K, x\in M_F, m\in M_E\cap K, a\in A_E\cap A(F), n\in N_E\cap N(F) \end{array} \right\}$$

in analogy with (6.16), treating k as an extra parameter which does not enter the differential equations. Since $\tilde{f}(kl,m) = \tilde{f}(k,lm)$ for $l \in M_F \cap K$ we also have $\beta^F_{E,\lambda}\tilde{f}(kl,m) = \beta^F_{E,\lambda}\tilde{f}(k,lm)$ and we can define the boundary value of f by

$$\beta^F_{E,\lambda} f(kman) = \beta^F_{E,\lambda}\tilde{f}(k,m)a^{\lambda-\rho}$$

for $k \in K$, $m \in M_F$, $a \in A_F$ and $n \in N_F$. Then $\beta^F_{E,\lambda} f$ is well defined and we have

$$\beta^F_{E,\lambda} : \mathcal{B}(F;\lambda) \longrightarrow \mathcal{B}(E;\lambda) .$$

It is then easily seen that $\beta^F_{E,\lambda}$ is the inverse of $c^E_\lambda(F)\mathcal{P}^F_E$ where $c^E_\lambda(F)$ is given by the expression (5.7), the product taken only over those $\alpha \in \Sigma^+_o$ for which $\alpha \in \langle F\rangle \setminus \langle E\rangle$.

6.5 Notes and further results

The material of this chapter is new and due to the author, with the following exceptions.

In the harmonic case (i.e., $\lambda = \rho$) , Poisson integrals for other boundary components have been studied by A. Korányi, who has proved several Fatou-type theorems for this case ([d] and [e] - notice the difference between our B_F and Korányi's $B(F)$, the latter denotes the full parabolic subgroup $P_F = M_F A_F N_F$). See also Johnson [b] Lemma 2.1. Also Stein [a] has results for all boundary components in the case $\lambda = \rho$. These Fatou theorems of Korányi and Stein are generalized to other values of λ in the author's note [f]. Theorem 6.1.5 is given in [f], but the main step of its proof goes back to Helgason and Johnson [a]. Lemma 6.1.6 is a generalization of Harish-Chandra [c] I Lemma 43.

The proof of Proposition 6.2.2 is from Harish-Chandra [c] II (Lemma 25). For $F = \emptyset$ Proposition 6.2.6 is similar to Kashiwara et al. [a] p. 33 (with a different proof). For related results, see Bruhat [a] and Helgason [c].

The boundary value maps $\beta_{F,\bar{\sigma},\lambda}$ of Section 6.3 were constructed by T. Oshima (Oshima and Matsuki [b] p. 38). Theorem 6.3.4 is related to Harish-Chandra's own refinements of his spherical function expansion (see [c] I Lemma 34 and II) - cf. also Trombi and Varadarajan [a] Theorem 2.11. For a quite different approach to the expansions of ϕ_λ, see van den Ban [a].

7. Semisimple symmetric spaces

In this chapter some basic properties of semisimple symmetric spaces are established, and a fundamental family of functions related to a semisimple symmetric space is defined. In the next chapter we shall use these results to study harmonic analysis on semisimple symmetric spaces.

7.1 The orbits of symmetric subgroups

Let G be a semisimple, connected Lie group with finite center, and let H be a closed subgroup.

Definition. The homogeneous space G/H is called a *semisimple symmetric space* if there is an involution σ (i.e., an involutive automorphism) of G such that H is the connected component containing the identity of the group G_σ of fixed points for σ. In this case H is called a *symmetric subgroup*.

Remark. Often one includes in the definition also the spaces G/H where H is not connected and satisfies $(G_\sigma)_o \subset H \subset G_\sigma$. In this book, H is always connected. (It follows from Proposition 7.1.2 below that G_σ has a finite number of connected components.)

Throughout this chapter we assume that G is noncompact. (If G is compact, then every symmetric space G/H is Riemannian.)

Examples a) G/G and G/K are semisimple symmetric spaces with σ respectively the identity map and $\sigma = \theta$.

b) Let $\operatorname{diag} : G \longrightarrow G \times G$ be the diagonal map $\operatorname{diag}(x) = (x,x)$, then its image $\operatorname{diag} G$ is a symmetric subgroup of $G \times G$, the involution being given by $\sigma(x,y) = (y,x)$. This example is of particular interest since it exhibits G itself as a symmetric space: $(x,y)\operatorname{diag} G \longrightarrow xy^{-1}$ is a diffeomorphism of $G \times G/\operatorname{diag} G$ onto G.

c) Let G_c be a connected real Lie group with Lie algebra $\mathfrak{g}_c$, and let K_c be the subgroup corresponding to $\mathfrak{k}_c$. Then G_c/K_c is a semisimple symmetric space. On the level of Lie algebras, the involution σ of $\mathfrak{g}_c$ is the complex linear extension of the Cartan involution of $\mathfrak{g}$.

d) Let $G = SL(2,\mathbb{R})$ and let σ be the involution given by

$$\sigma\begin{pmatrix} a & b \\ c & d \end{pmatrix} = \begin{pmatrix} a & -b \\ -c & d \end{pmatrix}$$

then H consists of the positive diagonal elements, and the space G/H can be identified with a one-sheeted hyperboloid in $\mathbb{R}^3$.

e) Let $G = SO_o(p,q)$, where $p \geq 1$ and $q \geq 2$ are integers. Let σ be the involution of G given by $\sigma g = I g I^{-1}$ $(g \in G)$, where I is the diagonal matrix all of whose diagonal elements are 1 except the last one, which is -1. Then $H = SO_o(p,q-1)$ and we can identify G/H with the hyperboloid in $\mathbb{R}^{p+q}$ whose equation is $-x_1^2 - \ldots - x_p^2 + x_{p+1}^2 + \ldots + x_{p+q}^2 = 1$. (Since $SL(2,\mathbb{R})/\{\pm 1\} \simeq SO_o(2,1)$ Example d is the special case $p = 1$, $q = 2$.)

For a complete list of all semisimple symmetric spaces, see Berger [a].

Assume G/H is a semisimple symmetric space. We will denote by σ also the involution of $\mathfrak{g}$ which is the differential of σ at e. Let $\mathfrak{h}$ be the Lie algebra of H and $\mathfrak{q}$ its orthocomplement in $\mathfrak{g}$. Then $\mathfrak{g} = \mathfrak{h} \oplus \mathfrak{q}$ is the decomposition of $\mathfrak{g}$ into eigenspaces for σ, and

$$(7.1) \qquad [\mathfrak{h},\mathfrak{q}] \subset \mathfrak{q}, \quad [\mathfrak{q},\mathfrak{q}] \subset \mathfrak{h}.$$

In fact, if 0 is the only ideal of $\mathfrak{g}$ in $\mathfrak{h}$, then it is easily seen that (7.1) holds with equalities.

<u>Proposition 7.1.1</u> <u>There is a Cartan involution</u> θ <u>of</u> $\mathfrak{g}$ <u>which commutes with</u> σ.

<u>Proof</u>: Let τ be an arbitrary Cartan involution of $\mathfrak{g}$, and extend τ and σ to antilinear automorphisms, also denoted τ and σ, of $\mathfrak{g}_c$. Let γ denote complex conjugation on $\mathfrak{g}_c$ with respect to $\mathfrak{g}$, then γ commutes with σ and τ. By Helgason[j]Chapter III Theorem 7.1 there is an automorphism φ of $\mathfrak{g}_c$ such that $\varphi\tau\varphi^{-1}$

commutes with σ. By the proof in loc. cit. φ is given by $\varphi = ((\sigma\tau)^2)^{1/4}$, and hence φ commutes with γ. Let θ be the restriction of $\varphi\tau\varphi^{-1}$ to $\mathfrak{g}$, then θ has the required properties.

$\square$

Henceforth we fix θ as in Proposition 7.1.1 and take over notation from the preceding chapters accordingly. Since σ and θ commute we have

$$(7.2) \qquad \mathfrak{g} = \mathfrak{k}\cap\mathfrak{h} \oplus \mathfrak{k}\cap\mathfrak{q} \oplus \mathfrak{p}\cap\mathfrak{h} \oplus \mathfrak{p}\cap\mathfrak{q}\,.$$

Let $\mathfrak{h}^o = \mathfrak{k}\cap\mathfrak{h} + \sqrt{-1}\,\mathfrak{p}\cap\mathfrak{h} \subset \mathfrak{h}_c$, then $\mathfrak{h}^o$ is a subalgebra of the compact form $\mathfrak{u} = \mathfrak{k} + \sqrt{-1}\,\mathfrak{p}$. It immediately follows that $\mathfrak{h}$ is reductive and $\mathfrak{h} = \mathfrak{k}\cap\mathfrak{h} + \mathfrak{p}\cap\mathfrak{h}$ is a Cartan decomposition of $\mathfrak{h}$. Then $H\cap K$ is a maximal compact subgroup of H, in particular $H\cap K$ is connected.

<u>Proposition 7.1.2</u> <u>The mapping</u>

$$\mathfrak{p}\cap\mathfrak{h} \times \mathfrak{p}\cap\mathfrak{q} \times K \ni (X,Y,k) \longrightarrow \exp X \exp Y k \in G$$

<u>is a real analytic diffeomorphism onto</u> G.

<u>Proof</u>: Clearly the map is real analytic. To prove that it is injective, let $g = \exp X \exp Y k \in G$ be given. We will show how to obtain X, Y and k from g. By $G = \exp\mathfrak{p}\, K$ we may assume $g = \exp S$ where $S \in \mathfrak{p}$. First we use that $g = \theta g^{-1} = k^{-1}\exp Y \exp X$ to eliminate k:

$$(7.3) \qquad \exp 2S = \exp X \exp 2Y \exp X$$

from which it follows that

$$(7.4) \qquad \exp 2\sigma S = \exp X \exp\text{-}2Y \exp X\,.$$

Eliminating Y from (7.3) and (7.4), we get

$$(7.5) \qquad \exp 2\sigma S = \exp 2X \exp\text{-}2S \exp 2X$$

which is equivalent to

$$(7.6) \qquad \exp\text{-}S \exp 2\sigma S \exp\text{-}S = (\exp\text{-}S \exp 2X \exp\text{-}S)^2\,.$$

Now, if we let $T \in \mathfrak{p}$ be given by

(7.7) $$\exp 2T = \exp -S \exp 2\sigma S \exp -S$$

then it follows from (7.6) that

(7.8) $$\exp 2X = \exp S \exp T \exp S \quad .$$

This shows that X is uniquely determined by g. But then (7.3) shows that also Y, and hence k, are uniquely determined. Thus the map is injective.

The preceding equations also indicate how to construct an inverse: If $S \in \mathfrak{p}$ is given we define T, X and Y in $\mathfrak{p}$ by (7.7), (7.8) and (7.3). We claim that then:

(7.9) $$X \in \mathfrak{p} \cap \mathfrak{h}$$

(7.10) $$Y \in \mathfrak{p} \cap \mathfrak{q}$$

(7.11) $$\exp -Y \exp -X \exp S \in K \; .$$

Obviously these three statements will imply that $\exp S$ can be expressed as the proposition prescribes, and since the determination of X and Y from S is real analytic, the proposition will follow from that.

To prove (7.9) we first notice that (7.7) and (7.8) imply (7.6) and hence (7.5), and that (7.5) and (7.3) imply (7.4). Applying σ to (7.5) and rearranging gives that

$$\exp 2\sigma S = \exp 2\sigma X \exp -2S \exp 2\sigma X \quad .$$

Since X was uniquely determined from (7.5) it follows then that $\sigma X = X$, which proves (7.9).

Using (7.9) we get when applying σ to (7.3) that

$$\exp 2\sigma S = \exp X \exp 2\sigma Y \exp X$$

which compared with (7.4) shows that $\sigma Y = -Y$. Hence (7.10) holds.

Finally, since (7.3) is equivalent to

$$\exp -Y \exp -X \exp S = \exp Y \exp X \exp -S$$

we see that $\theta(\exp -Y \exp -X \exp S) = \exp -Y \exp -X \exp S$ which shows (7.11). $\square$

We will now give a proposition, which describes the decomposition of G/H into K-orbits (or of G/K into H-orbits).

Let $\mathfrak{g}_o = \mathfrak{k} \cap \mathfrak{h} + \mathfrak{p} \cap \mathfrak{q}$, and let $G_o \subset G$ be the corresponding analytic subgroup of G. We notice that G_o is a symmetric subgroup, corresponding to the involution $\sigma\theta$. Let $\mathfrak{b}$ be a maximal abelian subspace of $\mathfrak{p} \cap \mathfrak{q}$, then $\mathfrak{b}$ is unique up to conjugacy by $K \cap H$. Let $\Sigma_o = \Sigma(\mathfrak{b}, \mathfrak{g}_o)$ denote the set of restricted roots of $\mathfrak{b}$ in $\mathfrak{g}_o$, and choose a positive system Σ_o^+. Let $\mathfrak{b}_o^+ \subset \mathfrak{b}$ be the corresponding Weyl chamber, let $B_o^+ = \exp \mathfrak{b}_o^+$, and let $\overline{B_o^+}$ denote the closure of B_o^+. Let $M_b = Z_{K \cap H}(\mathfrak{b})$ denote the centralizer of $\mathfrak{b}$ in $K \cap H$. For $x \in \overline{B_o^+}$ we denote by $Z_{K \cap H}(x)$ the centralizer of x in $K \cap H$, then $M_b \subset Z_{K \cap H}(x)$ with equality if and only if $x \in B_o^+$.

<u>Proposition 7.1.3</u> <u>The mapping</u>

$$K \times \overline{B_o^+} \times H \ni (k,x,h) \longrightarrow kxh \in G$$

<u>is surjective and has the fiber</u>

$$\{(kl^{-1}, x, lh) \mid l \in Z_{K \cap H}(x)\}$$

<u>above</u> kxh. <u>The mapping</u>

$$K/M_b \times B_o^+ \ni (kM_b, x) \longrightarrow kxH \in G/H$$

<u>is an analytic diffeomorphism onto a dense open set in</u> G/H.

<u>Proof</u>: This follows from the preceding proposition combined with the Cartan decomposition (3.3) of G_o and the decomposition $H = H \cap K \exp \mathfrak{p} \cap \mathfrak{h}$. □

Let P be a minimal parabolic subgroup of G. It is also of importance to describe the decomposition of G/P into orbits of H. However, this description is quite complicated and for this reason we will not give a complete description here (the complete result is given in Matsuki [a]).

We say that a maximal abelian subspace $\mathfrak{a}$ of $\mathfrak{p}$ is <u>$\mathfrak{h}$-maximal</u> if $\mathfrak{t} = \mathfrak{a} \cap \mathfrak{h}$ is a maximal abelian subspace of $\mathfrak{p} \cap \mathfrak{h}$. Then $\mathfrak{a}$ is invariant by σ, since $X \in \mathfrak{a}$ implies $X + \sigma X \in \mathfrak{t}$ by maximality, and hence $\sigma X \in \mathfrak{a}$. On the other hand, if

$\mathfrak{b} = \mathfrak{a} \cap \mathfrak{q}$ is maximal abelian in $\mathfrak{p} \cap \mathfrak{q}$, then we say that $\mathfrak{a}$ is $\mathfrak{q}$-maximal. We notice that $\mathfrak{a}$ is $\mathfrak{q}$-maximal if and only if it is $\mathfrak{g}_o$-maximal.

Lemma 7.1.4 Assume $\mathfrak{a}$ is $\mathfrak{h}$-maximal (resp. $\mathfrak{q}$-maximal) and let $\mathfrak{g}_n^{\mathfrak{t}}$ (resp. $\mathfrak{g}_n^{\mathfrak{b}}$) denote the noncompact semisimple part of the centralizer $\mathfrak{g}^{\mathfrak{t}}$ (resp. $\mathfrak{g}^{\mathfrak{b}}$) of $\mathfrak{t}$ (resp. $\mathfrak{b}$) in $\mathfrak{g}$. Then $\mathfrak{g}_n^{\mathfrak{t}} \subset \mathfrak{g}_o$ (resp. $\mathfrak{g}_n^{\mathfrak{b}} \subset \mathfrak{h}$).

Proof: By the maximality of $\mathfrak{t}$ we have $\mathfrak{g}_n^{\mathfrak{t}} \cap \mathfrak{h} \subset \mathfrak{g}_n^{\mathfrak{t}} \cap \mathfrak{k}$. Since $\mathfrak{g}_n^{\mathfrak{t}}$ is semisimple we have

$$\mathfrak{g}_n^{\mathfrak{t}} \cap \mathfrak{k} = [\,\mathfrak{g}_n^{\mathfrak{t}} \cap \mathfrak{p}\,,\ \mathfrak{g}_n^{\mathfrak{t}} \cap \mathfrak{p}\,]$$

and therefore it follows that $\mathfrak{g}_n^{\mathfrak{t}} \cap \mathfrak{k} = \mathfrak{g}_n^{\mathfrak{t}} \cap \mathfrak{h}$ and $\mathfrak{g}_n^{\mathfrak{t}} \cap \mathfrak{p} = \mathfrak{g}_n^{\mathfrak{t}} \cap \mathfrak{q}$. Hence $\mathfrak{g}_n^{\mathfrak{t}} \subset \mathfrak{g}_o$. The statement in parenthesis follows by exchanging $\mathfrak{h}$ with $\mathfrak{g}_o$. □

Lemma 7.1.5 All the maximal abelian subspaces of $\mathfrak{p}$ that are $\mathfrak{h}$-maximal (resp. $\mathfrak{q}$-maximal) are mutually $K \cap H$-conjugate.

Proof: Let $\mathfrak{a}$ and $\mathfrak{a}'$ be two maximal abelian subspaces of $\mathfrak{p}$ which are $\mathfrak{h}$-maximal. Then $\mathfrak{t} = \mathfrak{a} \cap \mathfrak{h}$ is conjugate to $\mathfrak{t}' = \mathfrak{a}' \cap \mathfrak{h}$ by $K \cap H$ since both $\mathfrak{t}$ and $\mathfrak{t}'$ are maximal abelian split subspaces for $\mathfrak{h}$. Thus we may assume $\mathfrak{t} = \mathfrak{t}'$. Then $\mathfrak{a}$ and $\mathfrak{a}'$ are maximal abelian split subspaces for $\mathfrak{g}^{\mathfrak{t}}$, and from Lemma 7.1.4 we get that they are mutually conjugate by $K \cap H$. □

Let $\mathfrak{a} \subset \mathfrak{p}$ be $\mathfrak{h}$-maximal, and let $\mathfrak{t} = \mathfrak{a} \cap \mathfrak{h}$. Let Σ^+ be a positive set for the root system Σ of $\mathfrak{a}$ in $\mathfrak{g}$. We say that Σ^+ is $\mathfrak{h}$-compatible if $\alpha \in \Sigma^+$ and $\alpha|_{\mathfrak{t}} \neq 0$ imply $\sigma\alpha \in \Sigma^+$. Let $\Sigma_h = \Sigma(\mathfrak{t}, \mathfrak{h})$ be the system of roots of $\mathfrak{t}$ in $\mathfrak{h}$, then

$$\Sigma_h^+ = \{\beta \in \Sigma_h \mid \exists \alpha \in \Sigma^+ : \beta = \alpha|_{\mathfrak{t}}\} \tag{7.12}$$

is a positive system for Σ_h.

Similarly, if $\mathfrak{a}$ is $\mathfrak{q}$-maximal, we say that Σ^+ is

$\mathfrak{q}$-compatible if $\alpha \in \Sigma^+$ and $\alpha|_{\mathfrak{b}} \neq 0$ imply $-\sigma\alpha \in \Sigma^+$, where $\mathfrak{b} = \mathfrak{a} \cap \mathfrak{q}$. Then Σ^+ is $\mathfrak{q}$-compatible if and only if it is $\mathfrak{g}_o$-compatible.

Let $\mathfrak{a}$ be any σ-stable maximal abelian subspace of $\mathfrak{p}$, and let $W \cong N_K(\mathfrak{a})/Z_K(\mathfrak{a})$ be the Weyl group of Σ. Then σ induces involutions of $N_K(\mathfrak{a})$ and $Z_K(\mathfrak{a})$, and hence of W. We have $(\sigma w)(H) = \sigma(w(\sigma H))$ for $H \in \mathfrak{a}$ and $w \in W$. Let $W_\sigma(\mathfrak{a})$ denote the set of fixed points in W for σ. We then have with $\mathfrak{t} = \mathfrak{a} \cap \mathfrak{h}$:

$$W_\sigma(\mathfrak{a}) = \{w \in W \mid w(\mathfrak{t}) = \mathfrak{t}\} \cong (N_K(\mathfrak{a}) \cap N_K(\mathfrak{t}))/Z_K(\mathfrak{a}) .$$

Notice that the subgroup $W_\sigma(\mathfrak{a})$ of W depends on $\mathfrak{a}$. When $\mathfrak{a}$ is fixed and confusion unlikely, we write W_σ for $W_\sigma(\mathfrak{a})$.

<u>Lemma 7.1.6</u> <u>Assume $\mathfrak{a}$ is $\mathfrak{h}$-maximal and let Σ^+ and $\Sigma^{+\sim}$ be two $\mathfrak{h}$-compatible positive systems for Σ. Then there exists $w \in W_\sigma(\mathfrak{a})$ such that $\Sigma^{+\sim} = w\Sigma^+$.</u>

<u>Proof</u>: The proof is postponed to the next section. □

<u>Proposition 7.1.7</u> <u>Assume $\mathfrak{a}$ is $\mathfrak{h}$-maximal (resp. $\mathfrak{q}$-maximal). Then $W_\sigma(\mathfrak{a})$ acts simply transitively on the set of $\mathfrak{h}$-compatible (resp. $\mathfrak{q}$-compatible) positive systems for Σ.</u>

<u>Proof</u>: The only non-trivial part of this proposition is Lemma 7.1.6.

Let $W_{K\cap H}(\mathfrak{a}) = \{w \in W \mid \exists k \in K \cap H : w = \mathrm{Ad}_{\mathfrak{a}}\, k\} \cong N_{K\cap H}(\mathfrak{a})/Z_{K\cap H}(\mathfrak{a})$ then $W_{K\cap H}(\mathfrak{a}) \subset W_\sigma(\mathfrak{a})$, and it follows from the preceding lemma that $W_\sigma(\mathfrak{a})/W_{K\cap H}(\mathfrak{a})$ acts simply transitively on the set of $K\cap H$ conjugacy classes of $\mathfrak{h}$-compatible (resp. $\mathfrak{q}$-compatible) positive systems for Σ.

Notice that if a positive system Σ_h^+ for Σ_h is fixed then every $K\cap H$ conjugacy class of $\mathfrak{h}$-compatible positive systems for Σ contains a representative Σ^+ satisfying (7.12).

<u>Proposition 7.1.8 (i)</u> <u>Assume $\mathfrak{a}$ is $\mathfrak{h}$-maximal and Σ^+ is $\mathfrak{h}$-compatible. Then the set HwP is closed for each $w \in W_\sigma(\mathfrak{a})$.</u>

(ii) Assume $\mathfrak{a}$ is $\mathfrak{q}$-maximal and Σ^+ is $\mathfrak{q}$-compatible. Then the set HwP is open for each $w \in W_\sigma(\mathfrak{a})$.

Proof: (i) Let $G = KAN$ be the Iwasawa decomposition of G corresponding to Σ^+. Then since Σ^+ is $\mathfrak{h}$-compatible, $H = (H \cap K)(H \cap A)(H \cap N)$ gives an Iwasawa decomposition of H. Therefore $HP = (H \cap K)P$, which is closed. Similarly we get by the compatibility of $w\Sigma^+$ that $HwP = (H \cap K)wP$ is closed for $w \in W_\sigma(\mathfrak{a})$.

(ii) Exchanging wPw^{-1} with P, we may assume $w = e$. To prove that HP is open it is enough to show that $\mathfrak{g} = \mathfrak{h} + \mathfrak{m} + \mathfrak{a} + \mathfrak{n}$. Since $\mathfrak{g} = \bar{\mathfrak{n}} + \mathfrak{m} + \mathfrak{a} + \mathfrak{n}$ it suffices to prove that $\bar{\mathfrak{n}} \subset \mathfrak{h} + \mathfrak{n}$. Let $\alpha \in \Sigma^+$ and let $X \in \mathfrak{g}^{-\alpha}$. If $\alpha|_{\mathfrak{b}} \neq 0$ then $-\sigma\alpha \in \Sigma^+$ by the $\mathfrak{q}$-compatibility of Σ^+, and hence $\sigma X \in \mathfrak{n}$. Thus $X = (X + \sigma X) - \sigma X \in \mathfrak{h} + \mathfrak{n}$. On the other hand, if $\alpha|_{\mathfrak{b}} = 0$ then $X \in \mathfrak{h}$ by Lemma 7.1.4. □

It remains to detect whether there are other closed (resp. open) orbits of H in G/P. In fact this is not the case, but we shall not need that result here, and therefore we refer to Matsuki [a] for the proof, as well as for the classification of the orbits that are neither closed nor open. In particular we conclude that the number of closed (resp. open) orbits of H in G/P is $|W_\sigma(\mathfrak{a})|/|W_{K\cap H}(\mathfrak{a})|$, where $\mathfrak{a}$ is $\mathfrak{h}$-maximal (resp. $\mathfrak{q}$-maximal) in $\mathfrak{p}$.

7.2 Root systems

We use the notation of the preceding section. Let $\mathfrak{a} \subset \mathfrak{p}$ be a maximal abelian subspace which is $\mathfrak{h}$-maximal, and let $\mathfrak{t} = \mathfrak{a} \cap \mathfrak{h}$. Let $\Sigma = \Sigma(\mathfrak{a}, \mathfrak{g})$ and let

$$R_t = \{\alpha|_{\mathfrak{t}} \mid \alpha \in \Sigma, \alpha|_{\mathfrak{t}} \neq 0\} \subset \mathfrak{t}^* .$$

It is our aim in this section to prove the following result:

Proposition 7.2.1 The set R_t is a root system on $\mathfrak{t}$, and its Weyl group is naturally identified with $N_K(\mathfrak{t})/Z_K(\mathfrak{t})$.

Proof: Let $\mu, \lambda \in R_t$ and put $N = \frac{2<\lambda,\mu>}{<\lambda,\lambda>}$. For R_t to be a root system on $\mathfrak{t}$ we must prove that $N \in \mathbb{Z}$ and $s_\lambda\mu = \mu - N\lambda \in R_t$ (it is obvious that R_t spans $\mathfrak{t}^*$).

Let $\mathfrak{g}^\lambda$ denote the weight space for λ, then $\mathfrak{g}^\lambda$ is invariant under σ since $\mathfrak{t} \subset \mathfrak{h}$. Therefore either $\mathfrak{g}^\lambda \cap \mathfrak{h} \neq 0$ or $\mathfrak{g}^\lambda \cap \mathfrak{q} \neq 0$. Pick X in one of these spaces, normalized to $B_\theta(X,X) = <\theta X, X> = -1$.

It follows easily that $[X,\theta X] \in \mathfrak{p} \cap \mathfrak{h}$, and since $[X,\theta X]$ commutes $\mathfrak{t}$ hence $[X,\theta X] \in \mathfrak{t}$.

We claim now that for all $\gamma \in \mathfrak{t}^*$:

$$\gamma([X,\theta X]) = -<\gamma,\lambda> . \tag{7.13}$$

To prove the claim, recall that $<\gamma,\lambda>$ by definition is $\lambda(H_\gamma)$ where $\gamma(H) = <H_\gamma, H>$ for all $H \in \mathfrak{t}$. Therefore:

$$\gamma([X,\theta X]) = <H_\gamma,[X,\theta X]> = <[H_\gamma,X],\theta X> = \lambda(H_\gamma)<X,\theta X>$$

and (7.13) follows.

Let $h = -(\frac{1}{2}<\lambda,\lambda>)^{-1}[X,\theta X]$, $x = (\frac{1}{2}<\lambda,\lambda>)^{-1/2}X$ and $y = -(\frac{1}{2}<\lambda,\lambda>)^{-1/2}\theta X$. Then using (7.13) we have that $[h,x] = 2x$, $[h,y] = -2y$ and $[x,y] = h$. Therefore we get a homomorphism φ : $sl(2,\mathbb{R}) \longrightarrow \mathfrak{g}$ by taking

$$\varphi\begin{pmatrix}0 & 1\\ 0 & 0\end{pmatrix} = x, \quad \varphi\begin{pmatrix}0 & 0\\ 1 & 0\end{pmatrix} = y, \quad \varphi\begin{pmatrix}1 & 0\\ 0 & -1\end{pmatrix} = h .$$

For $Z \in \mathfrak{g}^\mu$ we have $[h,Z] = \mu(h)Z = NZ$ by (7.13). Thus N is a weight of a finite dimensional representation of $sl(2,\mathbb{R})$, whence $N \in \mathbb{Z}$.

Define $k \in K$ by $k = \exp\frac{\pi}{2}(x-y)$, then it follows by computations in $SL(2,\mathbb{R})$ that $\mathrm{Ad}\,k(H) = s_\lambda H$ for $H \in \mathfrak{t}$, and hence $k \in N_K(\mathfrak{t})$ with $s_\lambda = \mathrm{Ad}\,k$. But then

$$\mathfrak{g}^{s_\lambda\mu} = \mathrm{Ad}\,k(\mathfrak{g}^\mu) \neq 0$$

so $s_\lambda\mu \in R_t$, and R_t is a root system.

Since the Weyl group of R_t is generated by the reflections we also get that it can be identified with a subgroup of $N_K(\mathfrak{t})/Z_K(\mathfrak{t})$. To complete the proof of the proposition, it suffices to show that $N_K(\mathfrak{t})/Z_K(\mathfrak{t})$ acts freely on the Weyl chambers for R_t, that is, if R_t^+ is a positive system for R_t and $\mathrm{Ad}\,k(R_t^+) = R_t^+$ for some $k \in N_K(\mathfrak{t})$ then $k \in Z_K(\mathfrak{t})$. For this we need a lemma.

Recall from the preceding section the group

$$W_\sigma = (N_K(\mathfrak{t}) \cap N_K(\mathfrak{a}))/Z_K(\mathfrak{a}) \quad .$$

Since $Z_K(\mathfrak{a}) \subset Z_K(\mathfrak{t})$ there is a natural map $w \to \tilde{w}$ of W_σ into $N_K(\mathfrak{t})/Z_K(\mathfrak{t})$. On $\mathfrak{t}$, $\tilde{w}$ acts as the restriction of w.

Let $W^{\mathfrak{t}}$ be the subgroup of W which leaves $\mathfrak{t}$ pointwise fixed, then $W^{\mathfrak{t}}$ is the Weyl group of the root system $\Sigma^{\mathfrak{t}} = \{\alpha \in \Sigma \mid \alpha|_{\mathfrak{t}} = 0)\}$, which is the set of restricted roots for $\mathfrak{a}$ in $\mathfrak{g}^{\mathfrak{t}}$. We have $W^{\mathfrak{t}} \subset W_\sigma$.

<u>Lemma 7.2.2</u> <u>The map</u> $w \to \tilde{w}$ <u>induces an exact sequence</u>

$$0 \to W^{\mathfrak{t}} \to W_\sigma \to N_K(\mathfrak{t})/Z_K(\mathfrak{t}) \to 0 \quad .$$

<u>Proof:</u> It is obvious that $W^{\mathfrak{t}}$ is the kernel of $w \to \tilde{w}$. Let $k \in N_K(\mathfrak{t})$. Then $\mathrm{Ad}\,k(\mathfrak{a})$ is conjugated to $\mathfrak{a}$ by $Z_{K \cap H}(\mathfrak{t})$, as follows from Lemma 7.1.4. Thus, modulo $Z_K(\mathfrak{t})$, k is contained in $N_K(\mathfrak{t})$, so $w \to \tilde{w}$ is surjective. □

Let $k \in N_K(\mathfrak{t})$ and assume $\mathrm{Ad}\,k(R_t^+) = R_t^+$. According to the lemma $kZ_K(\mathfrak{t}) = \tilde{w}$ for some $w \in W_\sigma$. Choose a positive system Σ^+ for Σ such that $\alpha|_{\mathfrak{t}} \in R_t^+ \cup \{0\}$ for $\alpha \in \Sigma^+$, and let $\Sigma^{\mathfrak{t}+} = \Sigma^{\mathfrak{t}} \cap \Sigma^+$. Then $w(\Sigma^{\mathfrak{t}+})$ is another positive system for $\Sigma^{\mathfrak{t}}$, so $w(\Sigma^{\mathfrak{t}+}) = w'(\Sigma^{\mathfrak{t}+})$ for some $w' \in W^{\mathfrak{t}}$. Since $w(R_t^+) = w'(R_t^+) = R_t^+$ we get that $w(\Sigma^+) = w'(\Sigma^+)$, and hence $w = w' \in W^{\mathfrak{t}}$. Therefore $\tilde{w}$ is trivial and $k \in Z_K(\mathfrak{t})$. This completes the proof of Proposition 7.2.1. □

<u>Proof of Lemma 7.1.6</u> Let Σ^+ and $\Sigma^{+\sim}$ be two $\mathfrak{h}$-compatible positive systems for Σ, and let R_t^+ and $R_t^{+\sim}$ be the corresponding positive systems for R_t obtained by restriction. By Proposition 7.2.1 and Lemma 7.2.2 there exists $w \in W_\sigma$ such that $R_t^{+\sim} = \tilde{w}R_t^+$, i.e., $\alpha \in \Sigma^+$ and $\alpha|_{\mathfrak{t}} \neq 0$ imply $w\alpha \in \Sigma^{+\sim}$. Conjugating w by an element of $W^{\mathfrak{t}}$ we then obtain that $w\Sigma^+ = \Sigma^{+\sim}$. □

Now let $\mathfrak{a} \subset \mathfrak{p}$ be $\mathfrak{q}$-maximal, and let

$$R_b = \{\alpha|_{\mathfrak{b}} \mid \alpha \in \Sigma, \ \alpha|_{\mathfrak{b}} \neq 0\} \subset \mathfrak{b}^*$$

then R_b is a root system with Weyl group $N_K(\mathfrak{b})/Z_K(\mathfrak{b})$, which follows by replacing $\mathfrak{h}$ with $\mathfrak{g}_o$. Choose Σ^+ $\mathfrak{q}$-compatibibly, let $R_b^+ = \{\alpha|_{\mathfrak{b}} \mid \alpha \in \Sigma^+, \ \alpha|_{\mathfrak{b}} \neq 0\}$, and let $n = \dim \mathfrak{a}$ and $\ell = \dim \mathfrak{b}$. Let Δ denote the set of simple roots for Σ^+ and let $\Delta^{\mathfrak{b}} = \{\alpha \in \Delta \mid \alpha|_{\mathfrak{b}} = 0\}$. Then $\Delta^{\mathfrak{b}}$ is the set of simple roots for $\Sigma^{\mathfrak{b}+} = \{\alpha \in \Sigma^+ \mid \alpha|_{\mathfrak{b}} = 0\}$.

<u>Lemma 7.2.3</u> <u>There is a permutation</u> $\alpha \to \alpha'$ <u>of order</u> 2 <u>of the set</u> $\Delta \setminus \Delta^{\mathfrak{b}}$ <u>such that</u>

$$\theta\sigma\alpha = \alpha' + \Sigma_{\beta \in \Delta^{\mathfrak{b}}} n(\alpha, \beta)\beta$$

<u>for all</u> $\alpha \in \Delta \setminus \Delta^{\mathfrak{b}}$, <u>for some</u> $n(\alpha, \beta) \in \mathbf{Z}_+$.

<u>Proof</u>: We have $\theta\sigma\alpha \in \Sigma^+$ by the $\mathfrak{q}$-compatibility, and hence there exist integers $n(\alpha, \beta) \in \mathbf{Z}_+$ such that

$$\theta\sigma\alpha = \Sigma_{\beta \in \Delta} n(\alpha, \beta)\beta .$$

Since $\theta\sigma\alpha|_{\mathfrak{b}} \neq 0$, $n(\alpha,\alpha') \neq 0$ for at least one $\alpha' \in \Delta \setminus \Delta^{\mathfrak{b}}$. Applying $\theta\sigma$ once more we get

$$\alpha = \Sigma_{\beta \in \Delta} n(\alpha, \beta)\theta\sigma\beta = \Sigma_{\beta \in \Delta} \Sigma_{\gamma \in \Delta} n(\alpha, \beta)n(\beta, \gamma)\gamma$$

showing that $\Sigma_{\beta \in \Delta} n(\alpha, \beta)n(\beta, \gamma) = \delta_{\alpha, \gamma}$. Since $n(\alpha, \beta) \geq 0$ for all α, β it easily follows that $n(\alpha, \alpha') = n(\alpha', \alpha) = 1$ and $n(\alpha, \beta) = n(\beta, \alpha) = 0$ for $\beta \in \Delta \setminus \Delta^{\mathfrak{b}}$, $\beta \neq \alpha'$. From this the lemma follows. □

Let ℓ_1 denote the number of roots $\alpha \in \Delta \setminus \Delta^{\mathfrak{b}}$ such that $\alpha \neq \alpha'$, then the preceding lemma implies

<u>Lemma 7.2.4</u> <u>The elements</u> $\alpha_1, \ldots, \alpha_n$ <u>of</u> Δ <u>can be enumerated in such a way that</u>

$$\begin{aligned} \alpha_j|_{\mathfrak{b}} &= \alpha_{j+\ell_1}|_{\mathfrak{b}} && \text{for} \quad \ell - \ell_1 < j \leq \ell \\ \alpha_j|_{\mathfrak{b}} &= 0 && \text{for} \quad \ell + \ell_1 < j \leq n . \end{aligned}$$

In particular, the elements $\alpha_1|_{\mathfrak{b}}, \ldots, \alpha_\ell|_{\mathfrak{b}}$ constitute the simple roots of R_b^+ .

Let $H_1,\ldots,H_n$ be the basis of $\mathfrak{a}$ dual to $\alpha_1,\ldots,\alpha_n$, and let

$$(7.14) \qquad Y_j = \begin{cases} H_j & \text{if } j \leq \ell - \ell_1 \\ H_j + H_{j+\ell_1} & \text{if } \ell - \ell_1 < j \leq \ell \end{cases}$$

then it follows that $Y_1,\ldots,Y_\ell$ is the basis of $\mathfrak{b}$ dual to $\alpha_1|_{\mathfrak{b}}, \ldots, \alpha_\ell|_{\mathfrak{b}}$.

7.3 A fundamental family of functions

Let G/H be a semisimple symmetric space and K a maximal compact subgroup of G invariant under the involution (cf. Proposition 7.1.1). In this section we will define a family of functions on G/K which, as we shall see in the next chapter, is fundamental for the L^2-harmonic analysis on semisimple symmetric spaces. The remaining sections of the present chapter are devoted to studying these functions, in particular their asymptotic behavior.

Using notation from Section 7.1, let $\mathfrak{a} \subset \mathfrak{p}$ be a maximal abelian subspace which is $\mathfrak{h}$-maximal, and let $\Sigma^+ \subset \mathfrak{a}^*$ be an $\mathfrak{h}$-compatible positive system for the roots of $\mathfrak{a}$ in $\mathfrak{g}$. For the purpose of Chapter 8 (which is to construct discrete series), we actually need the fundamental functions only in the special case where $\mathfrak{a} \subset \mathfrak{h}$. However, we feel that the functions are of importance also in the general case (for instance, for the construction of continuous series) and therefore we develop the theory without this assumption until Section 7.6. This also has the advantage that it illustrates the analogy with the theory of spherical functions (Section 5.3) - cf. Example a below.

For each $w \in W_\sigma = W_\sigma(\mathfrak{a})$ we define a distribution T_w on K/M by

$$T_w(\varphi) = \int_{K \cap H} \varphi(kw)\,dk$$

for $\varphi \in C^\infty(K/M)$. By a slight abuse of notation we have identified w with the corresponding element in K/M . Notice that T_w depends only on the right coset $W_{K \cap H}(\mathfrak{a})w$ in W_σ .

Let $\lambda \in \mathfrak{a}_c^*$, then under the isomorphism of $\mathcal{B}(K/M)$ with $\mathcal{B}(G/P;L_\lambda)$ the distribution T_w corresponds to the following distribution $T_{w,\lambda}$ on G:

(7.15) $$T_{w,\lambda}(\varphi) = \int_{K\cap H}\int_M\int_A\int_N \varphi(k\,w\,m\,a\,n)a^{\lambda+\rho}\,dk\,dm\,da\,dn\ .$$

Thus $T_{w,\lambda}$ is the distribution given by normalized invariant measure on the closed set HwP in G (cf. Proposition 7.1.8(i)).

We now define the <u>fundamental function</u> $\psi_{w,\lambda} = \psi^H_{w,\lambda}$ on G/K as the Poisson transform of $T_{w,\lambda}$:

$$\psi_{w,\lambda} = \mathcal{P}\,T_{w,\lambda} \in \mathcal{A}(G/K;\mathcal{M}_\lambda)\ .$$

It then follows that

(7.16) $$\psi_{w,\lambda}(gK) = \int_{K\cap H} \exp\langle -\lambda-\rho\,,\,H(g^{-1}kw)\rangle\,dk$$

for $g \in G$. In particular $\psi_{w,\lambda}(e) = 1$ and $\psi_{w,\lambda}(kx) = \psi_{w,\lambda}(x)$ for $k \in K\cap H$, $x \in G/K$.

Notice that if $\Sigma^{+\sim}$ is another $\mathfrak{h}$-compatible positive system then $\Sigma^{+\sim} = s\Sigma^+$ for some $s \in W_\sigma$ by Lemma 7.1.6, and it easily follows that the family of functions $\psi^\sim_{w,\lambda}$ defined as above, but with respect to $\Sigma^{+\sim}$ instead of Σ^+, is identical to the family $\psi_{w,\lambda}$. In fact

$$\psi^\sim_{w,\lambda} = \psi_{ws,s^{-1}\lambda}\ .$$

This often allows us, when studying $\psi_{w,\lambda}$, to make the convenient assumption that $w = e$, which just means that we exchange Σ^+ with $w\Sigma^+$ and λ with $w\lambda$. We write ψ_λ or ψ^H_λ for $\psi_{e,\lambda}$.

Notice also that by Lemma 7.1.5 the choice of $\mathfrak{a}$ is inessential for the definition of $\psi^H_{w,\lambda}$. Essentially we thus have defined $|W_\sigma|/|W_{K\cap H}(\mathfrak{a})|$ families of functions ψ_λ which are lumped together in the notation $\psi_{w,\lambda}$.

Let the positive system Σ^+_h for $\Sigma_h = \Sigma(\mathfrak{t},\mathfrak{h})$ be fixed by (7.12). From the remarks after Proposition 7.1.7 we have that for each coset $W_{K\cap H}(\mathfrak{a})w$ the representative $w \in W_\sigma$ can be chosen such that $w\Sigma^+$ is compatible with Σ^+_h. This we will always do.

Examples a) In both cases $H = K$ and $H = G$, ψ_λ^H is the spherical function ϕ_λ.

b) Let $G = SL(2,\mathbb{R})$ and $H = \{$positive diagonal matrices$\}$. Then $K \cap H = \{e\}$ and

$$\psi_{w,\lambda}(x) = \exp\langle -\lambda-\rho, H(x^{-1}w)\rangle$$

where w is plus or minus the identity matrix.

For more examples, see Section 7.7.

Let $\mathfrak{t} = \mathfrak{a} \cap \mathfrak{h}$ and define $\mu_{w,\lambda} \in \mathfrak{t}_c^*$ by

$$\langle \mu_{w,\lambda}, X\rangle = \langle w\lambda + w\rho - 2\rho_h, X\rangle$$

for $X \in \mathfrak{t}$, where $\rho = \frac{1}{2}\sum_{\alpha\in\Sigma^+}(\dim \mathfrak{g}^\alpha)\alpha \in \mathfrak{a}^*$ and $\rho_h = \frac{1}{2}\sum_{\alpha\in\Sigma_h^+}(\dim \mathfrak{h}^\alpha)\alpha \in \mathfrak{t}^*$. In particular, let $\mu_\lambda = \mu_{e,\lambda}$.

Lemma 7.3.1 For $x \in G$ and $h \in H$ we have

$$\psi_{w,\lambda}(h x K) = \int_{K\cap H} \exp\langle \mu_{w,\lambda}, H(hk)\rangle \exp\langle -\lambda-\rho, H(x^{-1}kw)\rangle\, dk .$$

Proof: We may assume $w = e$. From (3.6) applied to the group H we get

$$\begin{aligned}\psi_\lambda(h x K) &= \int_{K\cap H} \exp\langle -\lambda-\rho, H(x^{-1}h^{-1}k)\rangle\, dk \\ &= \int_{K\cap H} \exp\langle -\lambda-\rho, H(x^{-1}h^{-1}\kappa(hk))\rangle \exp\langle -2\rho_h, H(hk)\rangle\, dk \\ &= \int_{K\cap H} \exp\langle -\lambda-\rho, H(x^{-1}k)\rangle \exp\langle \mu_\lambda, H(hk)\rangle\, dk\end{aligned}$$

which proves the lemma. □

Lemma 7.3.2 For each $x \in G$, $\lambda \in \mathfrak{a}_c^*$, and $w \in W_\sigma$ the function

$$h(K\cap H) \longrightarrow \psi_{w,\lambda}(h^{-1}x K)$$

on $H/K\cap H$ is an eigenfunction for all $D \in \mathbb{D}(H/K\cap H)$ with the eigenvalue $\chi_{-\mu_{w,\lambda}-\rho_h}(D)$.

Proof: This is an immediate consequence of the preceding lemma. □

Another consequence of Lemma 7.3.1, as we shall now see, is that for certain singular values of λ the function $\psi_{w,\lambda}$ is H-finite (i.e., its translates $\psi_{w,\lambda}(h\cdot)$ $(h \in H)$ span a finite dimensional space). Moreover the H-type of $\psi_{w,\lambda}$ (i.e., the equivalence class of the representation of H on the spanned space) is determined by $\mu_{w,\lambda}$.

Let $M^+ = \{\mu \in \mathfrak{t}_c^* \mid \frac{\langle\mu,\beta\rangle}{\langle\beta,\beta\rangle} \in \mathbb{Z}_+ \text{ for all } \beta \in \Sigma_h^+\}$.

Proposition 7.3.3 If $\mu_{w,\lambda} \in M^+$, then $\psi_{w,\lambda}$ is H-finite and the H-type of $\psi_{w,\lambda}$ is irreducible with lowest weight $-\mu_{w,\lambda}$ on $\mathfrak{t}$.

In the special case of $H = G$, the proposition says that if $\nu \in \mathfrak{a}^*$ satisfies

$$\frac{\langle\nu,\alpha\rangle}{\langle\alpha,\alpha\rangle} \in \mathbb{Z}_+ \quad \text{for all} \quad \alpha \in \Sigma^+ \tag{7.17}$$

then the spherical function $\phi_{\nu+\rho}$ is G-finite of irreducible type with lowest weight $-\nu$. Let V be the finite dimensional space generated by $\phi_{\nu+\rho}$, and let δ_ν denote the contragradient representation of G on the dual space V^*. Then δ_ν has highest weight ν on $\mathfrak{a}$; let $v_\nu \in V^*$ be a weight vector with this weight. Let $u = \phi_{\nu+\rho} \in V$, then u is fixed by K. We have that $\langle u,v_\nu\rangle \neq 0$ since otherwise $\langle u,\delta_\nu(g)v_\nu\rangle = 0$ for all $g \in G$ by Iwasawa decomposition $G = KAN$. We normalize v_ν such that $\langle u,v_\nu\rangle = 1$. It then follows that for $x \in G$:

$$\exp\langle\nu, H(x)\rangle = \langle u, \delta_\nu(x)v_\nu\rangle \quad . \tag{7.18}$$

Thus $\exp\langle\nu, H(x)\rangle$ is a matrix coefficient of δ_ν.

Proof: In the special case mentioned above Proposition 7.3.3 follows from Theorem 7.3.4 stated below. We prove the proposition by reduction to that special case.

It follows from (7.18) applied to the group H that if $\mu = \mu_{w,\lambda} \in M^+$ then

$$\exp\langle\mu, H(h)\rangle = \langle u, \delta_\mu(h)v_\mu\rangle$$

where δ_μ is a finite dimensional irreducible representation of H

with highest weight μ . Let $\delta_\mu^\vee$ denote the contragradient representation, then we have

$$\exp\langle\mu, H(hk)\rangle = \langle\delta_\mu^\vee(h^{-1})u, \delta_\mu(k)v_\mu\rangle \quad .$$

From Lemma 7.3.1 it therefore follows that

$$(7.19) \quad \psi_{w,\lambda}(h \times K) = \langle\delta_\mu^\vee(h^{-1})u, \int_{K\cap H}\delta_\mu(k)v_\mu \exp\langle -\lambda-\rho, H(x^{-1}kw)\rangle dk\rangle \quad .$$

This shows that $\delta_\mu^\vee(h^{-1})u \longrightarrow \psi_{w,\lambda}(h\cdot)$ is a well defined H-homomorphism from $\delta_\mu^\vee$ to the space generated by $\psi_{w,\lambda}$. □

Theorem 7.3.4 Let G be reductive and let $\nu \in \mathfrak{a}_c^*$ with $\mathrm{Re}\langle\nu+\rho,\alpha\rangle \geq 0$ for all $\alpha \in \Sigma^+$. The spherical function $\phi_{\nu+\rho}$ is G-finite if and only if

$$\frac{\langle\nu,\alpha\rangle}{\langle\alpha,\alpha\rangle} \in \mathbb{Z}_+ \quad \text{for all} \quad \alpha \in \Sigma^+ \quad .$$

In this case the G-type of $\phi_{\nu+\rho}$ is irreducible with lowest weight $-\nu$ on $\mathfrak{a}$. All irreducible finite dimensional representations of G which are spherical (i.e., have K-finite vectors) arise in this way.

Proof: See Helgason [n] Chapter V, Theorem 4.1. □

Though we do not need it in the sequel, we finish this section by proving a result which generalizes the fact that $\phi_{s\nu} = \phi_\nu$ for $s \in W$ (cf. Lemma 5.3.1). Recall that $W^{\mathfrak{t}} \subset W$ was defined in Section 7.2 as the subgroup which leaves $\mathfrak{t}$ pointwise fixed.

Proposition 7.3.5 Let $w \in W_\sigma$ and $s \in W^{\mathfrak{t}}$. Then $\psi_{w,s\lambda} = \psi_{w,\lambda}$ for all $\lambda \in \mathfrak{a}_c^*$.

Proof: We may assume $w = e$. Since ψ_λ depends analytically on λ we may also assume $\mathrm{Re}\langle\lambda,\alpha\rangle > 0$ for all $\alpha \in \Sigma^+(s)$ (notation from Section 6.1), such that the integral

$$c_\lambda(s) = \int_{\overline{N}_s} \exp\langle -\lambda-\rho, H(\overline{n})\rangle d\overline{n}$$

converges (cf. Lemma 6.1.3).

Let $x \in G$. Using the Iwasawa decomposition of x it easily follows that

$$c_\lambda(s)\exp\langle -s\lambda-\rho, H(x)\rangle = \int_{\overline{N}_s} \exp\langle -\lambda-\rho, H(x\,s\,\overline{n})\rangle\, d\overline{n} .$$

We therefore have

$$(7.20) \quad c_\lambda(s)\psi_{s\lambda}(xK) = \int_{K\cap H}\int_{\overline{N}_s} \exp\langle -\lambda-\rho, H(x^{-1}k\,s\,\overline{n})\rangle\, d\overline{n}\, dk .$$

We now use that $s \in W^{\mathfrak{t}}$. This has two consequences. The first is that $s \in K\cap H$, since $W^{\mathfrak{t}}$ is the Weyl group of $\mathfrak{g}^{\mathfrak{t}}$, and $\mathfrak{g}^{\mathfrak{t}}_n \subset \mathfrak{g}_o$ by Lemma 7.1.4. Hence s vanishes in the integrand of (7.20). Secondly, $s \in W^{\mathfrak{t}}$ implies $\alpha|_{\mathfrak{t}} = 0$ for all $\alpha \in \Sigma^+(s)$ by the $\mathfrak{h}$-compatibility of Σ^+. Therefore $\overline{\mathfrak{n}}_s \subset \mathfrak{g}^{\mathfrak{t}}_n$, and hence $\kappa(\overline{n}) \in K\cap H$ for all $\overline{n} \in \overline{N}_s$. Inserting this into (7.20) gives

$$\begin{aligned} c_\lambda(s)\psi_{s\lambda}(xK) &= \int_{K\cap H}\int_{\overline{N}_s} \exp\langle -\lambda-\rho, H(x^{-1}k)+H(\overline{n})\rangle\, d\overline{n}\, dk \\ &= c_\lambda(s)\psi_\lambda(xK) \end{aligned}$$

and the proposition is proved since $c_\lambda(s) \neq 0$. □

7.4 A differential property

By their definition as Poisson integrals the functions $\psi_{w,\lambda}$ are eigenfunctions for $\mathbb{D}(G/K)$, acting from the right on $C^\infty(G/K)$. We also have the enveloping algebra $U(\mathfrak{g})$ acting on $C^\infty(G/K)$ from the left, and we can thus study how this acts on $\psi_{w,\lambda}$. This algebra is of course too big for any nonconstant function to be an eigenfunction. On the other hand, the center $Z(\mathfrak{g})$ of $U(\mathfrak{g})$ acts as scalars on any eigenfunction for $\mathbb{D}(G/K)$. We will now show that $\psi_{w,\lambda}$ is an eigenfunction for a bigger algebra than $Z(\mathfrak{g})$, namely the centralizer $U(\mathfrak{g})^H$ of H in $U(\mathfrak{g})$. In case of $H = K$, where ψ_λ is the spherical function ϕ_λ, this is obvious since $\phi_\lambda(x^{-1}) = \phi_{-\lambda}(x)$.

Let $\mathfrak{n} = \sum_{\alpha\in\Sigma^+} \mathfrak{g}^\alpha$ and let $\mathfrak{m} = \mathfrak{h}^{\mathfrak{a}}$ be the centralizer of $\mathfrak{a}$ in $\mathfrak{h}$. For simplicity of statement we consider only $w = e$.

Theorem 7.4.1 (i) There is a homomorphism

$$\xi : U(\mathfrak{g})^H \longrightarrow U(\mathfrak{a})$$

uniquely determined by the property

$$(7.21) \qquad u - \xi(u) \in (\mathfrak{h} \cap \mathfrak{h})_c U(\mathfrak{g}) + U(\mathfrak{g})(\mathfrak{m} + \mathfrak{n})_c$$

for all $u \in U(\mathfrak{g})^H$.

(ii) For each $\lambda \in \mathfrak{a}_c^*$ and $u \in U(\mathfrak{g})^H$ we have

$$(7.22) \qquad u\psi_\lambda = \xi(u)(-\lambda-\rho)\psi_\lambda \ .$$

Proof: (i) Let $\mathfrak{n}_1 \subset \mathfrak{n}$ denote the sum of all root spaces $\mathfrak{g}^\beta$ where $\beta \in R_t^+$, i.e., $\beta = \alpha|_{\mathfrak{t}}$ where $\alpha \in \Sigma^+$ and $\alpha|_{\mathfrak{t}} \neq 0$. Let $\mathfrak{n}^{\mathfrak{t}} \subset \mathfrak{n}$ denote the centralizer of $\mathfrak{t}$ in $\mathfrak{n}$, then $\mathfrak{n} = \mathfrak{n}_1 \oplus \mathfrak{n}^{\mathfrak{t}}$.

We have $\mathfrak{g} = \theta\,\mathfrak{n}_1 \oplus \mathfrak{g}^{\mathfrak{t}} \oplus \mathfrak{n}_1$, and by a standard argument (Bourbaki [a] VIII §6 no. 4) there exists for each $u \in U(\mathfrak{g})^{\mathfrak{t}}$ an element $u_1 \in U(\mathfrak{g}^{\mathfrak{t}})$ such that

$$u - u_1 \in U(\mathfrak{g})\,\mathfrak{n}_{1c} \ .$$

We claim now that $\mathfrak{g}^{\mathfrak{t}} = \mathfrak{h} \cap \mathfrak{h}^{\mathfrak{t}} \oplus \mathfrak{a} \oplus \mathfrak{m} \cap \mathfrak{q} \oplus \mathfrak{n}^{\mathfrak{t}}$. In fact, let $\mathfrak{g}_n^{\mathfrak{t}}$ be the noncompact semisimple part of $\mathfrak{g}^{\mathfrak{t}}$, then $\mathfrak{g}^{\mathfrak{t}} = \mathfrak{g}_n^{\mathfrak{t}} + \mathfrak{g}^{\mathfrak{a}}$ where $\mathfrak{g}^{\mathfrak{a}} = \mathfrak{m} + \mathfrak{a}$. Moreover by Lemma 7.1.4 $\mathfrak{g}_n^{\mathfrak{t}} \subset \mathfrak{h} \cap \mathfrak{h}^{\mathfrak{t}} \oplus \mathfrak{a} \oplus \mathfrak{n}^{\mathfrak{t}}$, which proves the claim. It follows that there exists $u_2 \in U(\mathfrak{a})$ such that

$$u_1 - u_2 \in (\mathfrak{h} \cap \mathfrak{h}^{\mathfrak{t}})_c\, U(\mathfrak{g}^{\mathfrak{t}}) + U(\mathfrak{g}^{\mathfrak{t}})(\mathfrak{m} + \mathfrak{n}^{\mathfrak{t}})_c \ .$$

Thus the existence of $u_2 = \xi(u) \in U(\mathfrak{a})$ satisfying (7.21) is proved. As we shall see below, (7.21) implies (7.22), which proves both the uniqueness of γ and that it is a homomorphism.

(ii) For $x \in G$ we define

$$f(x) = \exp\langle -\lambda-\rho, H(x)\rangle$$

then $\psi_\lambda(x) = \int_{K\cap H} f(x^{-1}k)\,dk$. For $u \in U(\mathfrak{g})^H$ it follows that

$$(u\psi_\lambda)(x) = \int_{K\cap H}(u_r f)(x^{-1}k)dk \ .$$

From the definition of f it immediately follows that $X_r f = 0$ for $X \in \mathfrak{m}+\mathfrak{n}$, and $X_r f = \langle -\lambda-\rho, X\rangle f$ for $X \in \mathfrak{a}$. On the other hand, for $X \in \mathfrak{h}\cap\mathfrak{k}$ and $v \in U(\mathfrak{g})$ we have

$$\int_{K\cap H}((Xv)_r f)(x^{-1}k)dk = 0$$

and (ii) follows from (7.21). □

Remark. It follows from the proof of (i) that actually

$$u - \xi(u) \in (\mathfrak{h}\cap\mathfrak{k}^{\mathfrak{t}})_c\ U(\mathfrak{g}) + U(\mathfrak{g})(\mathfrak{m}\cap\mathfrak{q} + \mathfrak{n})_c$$

where $\mathfrak{h}\cap\mathfrak{k}^{\mathfrak{t}}$ is the centralizer of $\mathfrak{t}$ in $\mathfrak{h}\cap\mathfrak{k}$. Moreover, it follows from Proposition 7.3.5 that if we define $\eta : U(\mathfrak{a}) \longrightarrow U(\mathfrak{a})$ as the automorphism generated by $\eta(X) = X-\rho(H)$, then $\eta\circ\xi(U(\mathfrak{g})^H) \subset U(\mathfrak{a})^{W^{\mathfrak{t}}}$.

7.5 Asymptotic expansions

We will now use the machinery of Section 2 to derive an asymptotic expansion of the function $\psi_\lambda = \psi^H_{e,\lambda}$ on G/K (there will be no restriction in taking $w = e$). We assume that λ satisfies Assumption $(A)^\emptyset$ (cf. Section 6.3), that is, for each $w \in W$ and $1 \leq j \leq n$

$$(7.23) \qquad \langle w\lambda-\lambda, H_j\rangle \notin \mathbb{Z} \quad \text{if} \quad wH_j \neq H_j \ .$$

(The assumption can be weakened slightly, cf., Remark 7.5.3).

For the definition of ψ_λ the space $\mathfrak{a}$ was chosen to be $\mathfrak{h}$-maximal. However, we shall need also a $\mathfrak{q}$-maximal maximal abelian subspace of $\mathfrak{p}$, and to distinguish it we call the latter space $\hat{\mathfrak{a}}$. Let $\Sigma^\wedge = \Sigma(\hat{\mathfrak{a}}, \mathfrak{g})$ and choose $\Sigma^{\wedge+}$ $\mathfrak{q}$-compatibly. There exists $c \in K$ such that $\hat{\mathfrak{a}} = \mathrm{Ad}\,c\,(\mathfrak{a})$ and $\Sigma^{\wedge+} = \mathrm{Ad}\,c\,(\Sigma^+)$. We denote by $H \longrightarrow H^\wedge$ and $Y \longrightarrow \hat{Y}$ the corresponding isomorphisms $\mathfrak{a}_c \longrightarrow \hat{\mathfrak{a}}_c$ and $\mathfrak{a}_c^* \longrightarrow \hat{\mathfrak{a}}_c^*$. Let $W^\wedge$ be the Weyl group of $\Sigma^\wedge$, then we also have an isomorphism $w \longrightarrow \hat{w}$ of W with $W^\wedge$.

We enumerate the simple roots $\hat{\alpha}_1,\dots,\hat{\alpha}_n$ for $\Sigma^{\wedge+}$ according to

Lemma 7.2.4, and define $Y_1^\wedge,\dots,Y_\ell^\wedge \in \mathfrak{b} = \hat{\mathfrak{a}} \cap \mathfrak{q}$ as in (7.14). With the notation defined above these correspond to elements $\alpha_1,\dots,\alpha_n \in \Sigma^+$ and $Y_1,\dots,Y_\ell \in \mathfrak{a}$.

For $y \in \mathbb{R}_+^\ell$ let $b_y \in B$ be defined by

$$b_y = \exp\left(-\sum_{j=1}^{\ell} \log y_j\, Y_j^\wedge\right) .$$

Let R_b denote the root system of $\mathfrak{b}$ in $\mathfrak{q}$ and let $R_b^+ = \{\alpha|_{\mathfrak{b}} \mid \alpha \in \Sigma^{\wedge +},\ \alpha|_{\mathfrak{b}} \neq 0\}$. Let $\mathfrak{b}^+$ be the corresponding positive Weyl chamber in $\mathfrak{b}$, then $y \to b_y$ gives a bijection of $]0,1]^\ell$ onto $\overline{B^+} = \exp \overline{\mathfrak{b}^+}$.

To study the asymptotic behavior of ψ_λ we consider ψ_λ on elements of the form $h\,b\,K$ in G/K where $h \in H$ and $b \in \overline{B_o^+}$ (cf. Proposition 7.1.3). Since our choice of $\Sigma^{\wedge +}$ was arbitrary it suffices to take $b \in \overline{B^+}$.

In analogy with Theorem 6.3.4 we want an expression for the behavior of $\psi_\lambda(h\,b\,K)$ as $\log b$ tends to infinity in $\mathfrak{b}^+$, possibly "near the walls". With $b = b_y$ this means that y_j tends to 0 for some j and perhaps stays away from 0 for some other j. Let $\Theta \subset \{1,\dots,\ell\}$. We will thus study the behavior of $\psi_\lambda(h\,b_y K)$ when $y_j \to 0$ for $j \notin \Theta$. We may assume $\Theta \neq \{1,\dots,\ell\}$.

Let $W_\Theta = \{w \in W \mid w Y_j = Y_j$ for $j \notin \Theta\}$ and $W_\Theta^\wedge = \{\hat{w} \in W^\wedge \mid w \in W_\Theta\}$. As usual $W_\Theta \backslash W$ denotes the set of right cosets $\overline{w} = W_\Theta w$ in W. We notice that for $\overline{w} \in W_\Theta \backslash W$ the quantity $\langle w\lambda, Y_j\rangle$ makes sense for $\lambda \in \mathfrak{a}_c^*$ and $j \notin \Theta$.

<u>Theorem 7.5.1</u> <u>For each</u> $\lambda \in \mathfrak{a}_c^*$ <u>satisfying (7.23) and each</u> $\Theta \subset \{1,\dots,\ell\}$ <u>there exists an open set</u> $\Omega_\Theta \subset \mathbb{R}^\ell$ <u>containing</u>

$$\{y \mid y_i > 0 \ \text{if}\ i \in \Theta,\ y_j = 0 \ \text{if}\ j \notin \Theta\}$$

<u>and for each</u> $\overline{w} \in W_\Theta \backslash W$ <u>an analytic function</u> $\varphi_{\Theta,\overline{w}}^\lambda(h,y)$ <u>on</u> $H \times \Omega_\Theta$ <u>such that</u>

$$(7.24) \qquad \psi_\lambda(h\,b_y K) = \sum_{\overline{w} \in W_\Theta \backslash W} \varphi_{\Theta,\overline{w}}^\lambda(h,y) \prod_{j \notin \Theta} y_j^{\langle \rho - w\lambda, Y_j\rangle}$$

<u>for all</u> $h \in H$, $y \in \Omega_\Theta \cap \mathbb{R}_+^\ell$. $\varphi_{\Theta,\overline{w}}^\lambda$ <u>depends holomorphically on</u> λ <u>in</u> $\{\lambda \in \mathfrak{a}_c^* \mid (7.23)\text{ holds}\}$.

Proof: For the moment, let $E \subset \Delta$ be arbitrary and let $w \in W$. Let $P_E^\wedge = cP_E c^{-1}$, $B_E^\wedge = cB_E c^{-1}$ etc., and for $\lambda \in \mathfrak{a}_c^*$ let

$$\mathcal{B}(E;\lambda)^\wedge = \{f \in \mathcal{B}(G) \mid f(\cdot c^{-1}) \in \mathcal{B}(E;\lambda)\} .$$

Using $\hat{\mathfrak{a}}$ instead of $\mathfrak{a}$ we have from (6.16) the boundary value map under the assumption $(A)^E$:

$$\beta_{E,w\lambda}^\wedge : \mathcal{A}(G/K\ ;\mathcal{M}_\lambda) \longrightarrow \mathcal{B}(E;w\lambda)^\wedge .$$

Notice that from Proposition 7.1.8 we have that the subset $H\,P_E^\wedge$ of G is open. We will first prove

Lemma 7.5.2 $\beta_{E,w\lambda}^\wedge(\psi_\lambda)$ is analytic on $H\,P_E^\wedge$.

Proof: Let $f(h,p) = \beta_{E,w\lambda}^\wedge(\psi_\lambda)(h^{-1}p)$ for $h \in H$ and $p \in P_E^\wedge$. We claim that f is analytic in both variables. For the first variable this follows from Lemma 7.3.2 because $\beta_{E,w\lambda}^\wedge$ is a G-map. For the second variable this is obvious since $\beta_{E,w\lambda}^\wedge(\psi_\lambda) \in \mathcal{B}(E;w\lambda)^\wedge$. □

Consider the local coordinates given by φ_g $(g \in G)$ on $\tilde{X}$. Let

$$\hat{a}_t = \exp\,(-\Sigma_{t_j \neq 0} \log t_j\, \hat{H}_j)$$

for $t \in \mathbb{R}^n$. Using Lemma 7.5.2 and Theorem 2.5.6 as in the proofs of Theorems 5.3.2 and 6.3.4 we get the existence of an open set $\Omega \subset \mathbb{R}^n$ containing

$$\{t \in \mathbb{R}^n \mid t_i > 0 \text{ if } \alpha_i \in E,\ t_j = 0 \text{ if } \alpha_j \notin E\}$$

and analytic functions $\varphi_{\overline{w},g}^\lambda(\overline{n},t)$ on

$$\{(\overline{n},t) \in \overline{N} \times \Omega \mid g\,n\,\hat{a}_t \in H\,P_E^\wedge\}$$

for λ satisfying $(A)^E$, such that

$$(7.25)\quad \psi_\lambda(g\,\overline{n}\,\hat{a}_t K) = \sum_{\overline{w} \in W_E \backslash W} \varphi_{\overline{w},g}^\lambda(\overline{n},t) \prod_{\alpha_j \notin E} t_j^{<\rho - w\lambda, H_j>}$$

for $g\,\overline{n}\,\hat{a}_t \in H\,P_E^\wedge$ and $t \in \mathbb{R}_+^n \cap \Omega$. By 2.5.2 $\varphi_{\overline{w},g}^\lambda$ depends holomorphically on λ where $(A)^E$ holds.

To prove (7.24) take

$$(7.26)\qquad E=\{\alpha_j \mid j\le \ell \text{ and } j\in\Theta\}\cup\{\alpha_j \mid \ell<j\le \ell+\ell_1 \text{ and } j-\ell_1\in\Theta\}\cup \{\alpha_j \mid \ell+\ell_1<j\le n\} .$$

Let $g=h$, $n=e$ and

$$t_j=\begin{cases} y_j & \text{if } j\le \ell \\ y_{j-\ell_1} & \text{if } \ell<j\le \ell+\ell_1 \\ 1 & \text{if } \ell+\ell_1<j\le n .\end{cases}$$

Then $h\,b_y = g\,n\,\hat{a}_t \in H P_E^{\wedge}$ and

$$\prod_{\alpha_j\notin E} t_j^{<\rho-w\lambda,H_j>} = \prod_{j\notin\Theta} y_j^{<\rho-w\lambda,Y_j>} .$$

Since $W_E\subset W_\Theta$, (7.25) implies (7.24). □

<u>Remark 7.5.3</u> We have stated Theorem 7.5.1 under the condition $(A)^{\emptyset}$ but the proof shows that it actually holds for the given Θ under the condition $(A)^E$ where E is given by (7.26). This is a slightly weaker condition.

7.6 <u>The case of equal rank</u>

In this section we assume that the rank of G/K equals that of $H/H\cap K$. In other words, the assumption is that the $\mathfrak{h}$ -maximal space $\mathfrak{a}$ is in fact contained in $\mathfrak{h}$. We will show that this enables us to refine the asymptotic expansion of Theorem 7.5.1 such that some of the terms vanish. This case is thus extremely different from the special case of $H=K$, where none of the terms vanish ($c_{s\lambda}\neq 0$ for all $s\in W$) .

Since $\mathfrak{a}\subset\mathfrak{h}$ we have $\Sigma_h\subset\Sigma$, $W_\sigma(\mathfrak{a})=W$ and $W_{K\cap H}(\mathfrak{a})=W_h$, the Weyl group of Σ_h . Fix a positive system Σ_h^+ for Σ_h , then the elements of W/W_h correspond to the various choices of Σ^+ such that $\Sigma_h^+\subset\Sigma^+$. Thus we have one family of functions ψ_λ for each choice of positive system Σ^+ containing Σ_h^+ . Fix one such choice Σ^+ .

Let $\mathfrak{a}^*_+ = \{\lambda \in \mathfrak{a}^* \mid \langle\lambda,\alpha\rangle > 0, \quad \forall\alpha \in \Sigma^+\}$ be the positive Weyl chamber for Σ^+ in $\mathfrak{a}^*$, and let ${}^+\mathfrak{a} \subset \mathfrak{a}$ be the dual cone, that is

$${}^+\mathfrak{a} = \{H \in \mathfrak{a} \mid \lambda(H) \geq 0, \quad \forall\lambda \in \mathfrak{a}^*_+\} .$$

Then $\mathfrak{a}^+ \subset {}^+\mathfrak{a}$ (cf. Helgason [j] Chapter VII Lemma 2.20). From Theorem 7.3.4 we have that an element $H \in \mathfrak{a}$ is contained in ${}^+\mathfrak{a}$ if and only if $\eta(H) \geq 0$ for all highest weights η of spherical finite dimensional representations of G.

<u>Lemma 7.6.1</u> <u>If</u> $\mathfrak{a} \subset \mathfrak{h}$ <u>then</u> $H(x) \in {}^+\mathfrak{a}$ <u>for all</u> $x \in G_o$.

<u>Proof</u>: Let $x \in G_o$. We may assume that $x = \exp X$ for some $X \in \mathfrak{p} \cap \mathfrak{q}$. Let π be a spherical finite dimensional representation of G with highest weight $\eta \in \mathfrak{a}^*$. We claim that $\eta(H(x)) \geq 0$. Let π also denote the differentiated representation of $\mathfrak{g}_{\mathbb{C}}$, which we may assume is unitary on $\mathfrak{k} + \sqrt{-1}\,\mathfrak{p}$. Let v be a highest weight vector for π of norm 1. Then

$$\eta(H(x)) = \log \|\pi(x)v\|$$

for $x \in G$, and hence the claim is that $\|\pi(x)v\| \geq 1$. Let $f(t) = \|\pi(\exp tX)v\|^2 = \exp\langle 2\eta, H(\exp tX)\rangle$ for $t \in \mathbb{R}$.

We can expand v into an orthogonal sum of eigenvectors of $\pi(X)$: $v = \sum_{a \in \mathbb{R}} v_a$ (finite sum) where $\pi(X)v_a = av_a$. We then see that $f(t) = \sum_a e^{2at} \|v_a\|^2$.

On the other hand f is even, as follows from $H(\sigma g) = H(g)$ for $g \in G$. This in turn follows from the assumption that $\mathfrak{a} \subset \mathfrak{h}$, since σ then preserves K, A and N in the Iwasawa decomposition.

Therefore $f(t) = \sum_a \cosh(2at) \|v_a\|^2 \geq 1$. □

<u>Proposition 7.6.2</u> <u>If</u> $\mathfrak{a} \subset \mathfrak{h}$ <u>and</u> $\operatorname{Re}\lambda + \rho \in \overline{\mathfrak{a}^*_+}$ <u>then</u> ψ_λ <u>is bounded on</u> LBK <u>for every compact set</u> $L \subset H$.

<u>Proof</u>: By the preceding lemma $|\exp\langle -\lambda-\rho, H(x^{-1}k)\rangle| \leq 1$ for $x \in B$ and $k \in K \cap H$. The proposition then follows from Lemma 7.3.1. □

Using this boundedness we can now prove that some of the terms in the expansion of ψ_λ must vanish. Let $\Theta \subset \{1,\ldots,\ell\}$ be a proper subset and put

$$^{+}W(\Theta) = \{\bar{w} \in W_\Theta \backslash W \mid \forall j \notin \Theta : w^{-1}Y_j \in -{}^{+}\mathfrak{a}\} .$$

Note that $\bar{e} \notin {}^{+}W(\Theta)$ because $Y_j \in \mathfrak{a}^{+} \subset {}^{+}\mathfrak{a}$.

<u>Theorem 7.6.3</u> <u>Under the assumption that</u> $\mathfrak{a} \subset \mathfrak{h}$, <u>the coefficients</u> $\varphi^{\lambda}_{\Theta,\bar{w}}(h,y)$ <u>in the expansion (7.24) of Theorem 7.5.1 vanish identically for</u> $\bar{w}$ <u>outside the set</u> $^{+}W(\Theta)$.

<u>Proof:</u> Let $\bar{w}_o \notin {}^{+}W(\Theta)$. Then there exists $\lambda_o \in \mathfrak{a}^{*}_{+}$ and $i \notin \Theta$ such that $\langle w_o\lambda_o, Y_i\rangle > 0$. Since $\mathfrak{a}^{*}_{+}$ is a cone we may even assume $\langle w_o\lambda_o - \rho, Y_i\rangle > 0$. Thus there is an open set of λ's in $\mathfrak{a}^{*}_{c}$ for which $\mathrm{Re}\,\lambda \in \mathfrak{a}^{*}_{+}$, $\mathrm{Re}\langle w_o\lambda - \rho, Y_i\rangle > 0$, and (7.23) holds. By the holomorphic dependence on λ it suffices to show that $\varphi^{\lambda}_{\Theta,\bar{w}_o} = 0$ for λ in this set.

Let $\Xi = \{1,\dots,\ell\} \backslash \{i\}$. Then we have

$$\psi_\lambda(hb_yK) = \sum_{\bar{\sigma} \in W_\Xi\backslash W} \varphi^{\lambda}_{\Xi,\bar{\sigma}}(h,y)\, y_i^{\langle\rho-\sigma\lambda,Y_i\rangle} \tag{7.27}$$

for $h \in H$ and $y \in \Omega_\Xi \cap \mathbb{R}^{\ell}_{+}$. Since the functions $y_i^{\langle\rho-\sigma\lambda,Y_i\rangle}$ are linearly independent the expression (7.27) is unique. Comparing (7.27) with (7.24) we then get for $\bar{\sigma} \in W_\Xi\backslash W$, $h \in H$ and $y \in \Omega_\Xi \cap \Omega_\Theta \cap \mathbb{R}^{\ell}_{+}$ that

$$\varphi^{\lambda}_{\Xi,\bar{\sigma}}(h,y) = \sum \varphi^{\lambda}_{\Theta,\bar{w}}(h,y) \prod_{j \in \Xi\backslash\Theta} y_j^{\langle\rho-w\lambda,Y_j\rangle} \tag{7.28}$$

where the sum ranges over those $\bar{w} \in W_\Theta\backslash W$ for which $\sigma^{-1}Y_i = w^{-1}Y_i$.

In particular we have (7.28) for $\bar{\sigma} = W_\Xi w_o$, and since (7.28) is also unique it suffices to prove that $\varphi^{\lambda}_{\Xi,\bar{w}_o} = 0$. We have thus reduced the proof to the case where $\Delta\backslash\Theta$ has only one element, and may assume $\Theta = \Xi$.

Since $y_i^{\langle\rho-w_o\lambda,Y_i\rangle}$ is not bounded as $y_i \to 0$ by the assumption on λ whereas ψ_λ is bounded by Proposition 7.6.2, it follows from (7.27) that $\varphi^{\lambda}_{\Theta,\bar{w}_o}$ vanishes at $y_i = 0$. But then $\varphi^{\lambda}_{\Theta,\bar{w}_o} = 0$ identically by the last statement of Theorem 2.5.6. □

<u>Corollary 7.6.4</u> <u>If</u> $\mathrm{Re}\,\lambda \in \mathfrak{a}^{*}_{+}$ <u>and</u> $\mathfrak{a} \subset \mathfrak{h}$ <u>then for each compact set</u> $L \subset H$ <u>there exists a constant</u> $C_L > 0$ <u>and</u> $\varepsilon > 0$ <u>(both depending on</u> λ <u>)</u> <u>such that</u>

(7.29) $$|\psi_\lambda(h \exp YK)| \le C_L \exp(-(1+\varepsilon)\langle\hat{\rho},Y\rangle)$$

for all $h \in L$ and $Y \in \overline{\mathfrak{b}^+}$.

Proof: From Lemma 7.3.1 it follows that

$$|\psi_\lambda(h \exp YK)| \le C_L' \psi_{\mathrm{Re}\,\lambda}(\exp YK)$$

for $h \in L$, $Y \in \mathfrak{b}$ for some constant $C_L' > 0$. Therefore it suffices to prove (7.29) with $h = e$ and $\lambda \in \mathfrak{a}_+^*$.

If (7.29) holds for some $\lambda \in \mathfrak{a}_+^*$ then it holds for all $\lambda' \in \lambda + \mathfrak{a}_+^*$ because

$$\psi_{\lambda'}(xK) = \int_{K\cap H} \exp\langle\lambda-\lambda', H(x^{-1}k)\rangle \exp\langle-\lambda-\rho, H(x^{-1}k)\rangle\, dk$$

and $\exp\langle\lambda-\lambda', H(x^{-1}k)\rangle \le 1$ for $x \in B$ by Lemma 7.6.1. We may therefore assume (7.23) since there is always a smaller λ in $\mathfrak{a}_+^*$ satisfying this assumption. This allows us to use the asymptotic expansion (7.24) to derive the upper bound.

From Theorem 7.6.3 it follows that for each $\Theta \subset \{1,\dots,\ell\}$ the only nonzero summands of (7.24) are given by those $\overline{w} \in W_\Theta \backslash W$ for which $w\lambda(Y_j) < 0$ for all $j \notin \Theta$.

Choose $\varepsilon > 0$ such that

$$w\lambda(Y_j) \le -\varepsilon\rho(Y_j)$$

for all $j = 1,\dots,\ell$ and $w \in W$ with $w\lambda(Y_j) < 0$. We claim that then

$$\psi_\lambda(b_y K) \le C \prod_{j=1}^{\ell} y_j^{(1+\varepsilon)\rho(Y_j)}$$

for some constant $C > 0$. Obviously this claim implies (7.29).

Choose $\delta_1 > 0$ such that the set $K_\emptyset =]0,\delta_1]^\ell$ is contained in $\Omega_\emptyset$. Next choose $\delta_2 > 0$ such that $\delta_2 \le \delta_1$ and for each $i = 1,\dots,\ell$ the set

$$K_{\{i\}} = \{y \in \mathbb{R}^\ell \mid y_i \in [\delta_1, 1],\ y_j \in]0,\delta_2]\ j \ne i\}$$

is contained in $\Omega_{\{i\}}$. Proceeding like this we get a sequence of numbers

$$\delta_1 \ge \delta_2 \ge \dots \ge \delta_\ell > 0$$

such that for every subset Θ of $\{1,\dots,\ell\}$ with k elements $(k < \ell)$, the set

$$K_\Theta = \{y \in \mathbb{R}^\ell \mid y_i \in [\delta_k, 1] \ (i \in \Theta),\ y_j \in]0, \delta_{k+1}] \ (j \notin \Theta)\}$$

is contained in Ω_Θ. Finally we let $K_{\{1,\dots,\ell\}} = [\delta_\ell, 1]^\ell$. It easily follows that $\cup_\Theta K_\Theta =]0,1]^\ell$ (cf. Figure 7.1).

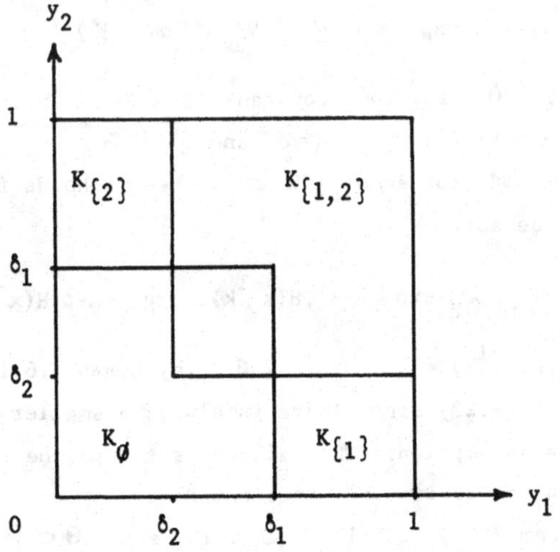

Figure 7.1

Let

$$C_1 = \sup\{|\varphi^\lambda_{\Theta,w}(h,y)| \mid h \in L,\ y \in K_\Theta,\ \Theta \subset \{1,\dots,\ell\},\ \overline{w} \in W_\Theta \setminus W\}.$$

Then for each $\Theta \subset \{1,\dots,\ell\}$ we get

$$\psi_\lambda(b_y K) \le |W_\Theta \setminus W|\ C_1 \prod_{j \notin \Theta} y_j^{(1+\varepsilon)\rho(Y_j)}$$

for all $y \in K_\Theta$, and the claim follows. $\square$

<u>Remark</u>: It is of importance to notice that the upper bound for ψ_λ given by (7.29) holds for all $\lambda \in \mathfrak{a}_+^*$ - also those for which the asymptotic expansion fails (i.e., (7.23) is not necessary for the upper bound).

7.7 <u>Examples</u>

We will compute the functions ψ_λ^H explicitly in two cases, one where G/K has rank one, and one of rank two. In both cases we use the explicit expression to derive the asymptotic expansion.

7.7.1 $G/H = SO_o(n,1)/SO(k) \times SO_o(n-k,1)$

Let G be the Lorentz group $G = SO_o(n,1)$ $(n \geq 2)$ and H the subgroup

$$H = \left\{ \begin{pmatrix} A & 0 \\ 0 & B \end{pmatrix} \middle| \; A \in SO(k)\,,\; B \in SO_o(n-k,1) \right\}$$

where $0 < k < n$ ($n = 2$ is Example b of Section 7.3).

We define the subalgebras $\mathfrak{h}$, $\mathfrak{a}$ and $\mathfrak{n}$ of $\mathfrak{g}$ as follows: Let E_{ij} be the matrix whose (i,j)-component equals 1 and all other components are 0. Put $X_{ij} = E_{ij} - E_{ji}$ for $i \neq j$, and $Y_i = E_{i,n+1} + E_{n+1,i}$ for $1 \leq i \leq n$. Then $\mathfrak{h}$ is defined as the span of all X_{ij} $(1 \leq i < j \leq n)$, $\mathfrak{a}$ as the span of Y_n, and $\mathfrak{n}$ as the span of all $Y_i - X_{in}$ $(1 \leq i \leq n-1)$. Then $\Sigma^+ = \{\alpha\}$ where $\alpha(Y_n) = 1$ and α has multiplicity $n-1$. We see that $\mathfrak{a} \subset \mathfrak{h}$, and that the multiplicity of α in Σ_h^+ is $n-k-1$. When $k < n-1$ there is thus only one family of functions ψ_λ, whereas when $k = n-1$ there are two such families $\psi_{\pm 1,\lambda}$.

As a maximal abelian subspace of $\mathfrak{p} \cap \mathfrak{q}$ we take $\mathfrak{b} = \mathbb{R}Y_1$. We then see that every element ℓ of $K \cap H = SO(k) \times SO(n-k)$ can be written as the product $\ell = \ell_a \ell_b$ of elements $\ell_a \in SO(k)$ and $\ell_b \in SO(n-k)$. From Lemma 7.3.1 it therefore follows that

$$\psi_\lambda(h^{-1}xK) = \int_{K \cap H} \exp\langle \mu_\lambda, H(h^{-1}\ell)\rangle\, dk \; \exp\langle -\lambda-\rho, H(x^{-1})\rangle$$

for $h \in H$ and $x \in B$, because $H(x^{-1}\ell) = H(\ell_b x^{-1}\ell_a) = H(x^{-1})$. The first factor in this expression is the spherical function $\phi_{-\mu_\lambda-\rho_H}$ on $H/H \cap K \simeq SO_o(n-k,1)/SO(n-k)$ and is thus explicitly known (essentially as a hypergeometric function). The second factor, which governs the asymptotic behavior on B, is easily computed as follows.

Let $u \in \mathbb{R}^{n+1}$ be the vector

$$u = \frac{1}{\sqrt{2}} \begin{bmatrix} 0 \\ \vdots \\ 0 \\ 1 \\ 1 \end{bmatrix}$$

then with the natural action of G on $\mathbb{R}^{n+1}$ we have $H(g) = (\log \|gu\|)Y_n$ for $g \in G$. If $\lambda \in \mathfrak{a}_c^*$ is given by

$\lambda(Y_n) = a \in \mathbb{C}$ and $x_t \in B$ by $x_t = \exp tY_1$ $(t \in \mathbb{R})$, then it follows that $\exp\langle -\lambda-\rho, H(x_t^{-1})\rangle$ equals $\cosh t$ raised to the power $-a - \frac{n-1}{2}$.

With $b_y = x_{-\log y}$ we thus have

$$\psi_\lambda(h\, b_y K) = \phi_{\mu_\lambda + \rho_H}(h) \left(\frac{y + y^{-1}}{2}\right)^{-a-\frac{n-1}{2}} \tag{7.30}$$

or

$$\psi_\lambda(h\, b_y K) = \varphi^\lambda(h, y)\, y^{a+\frac{n-1}{2}} \tag{7.31}$$

where $\varphi^\lambda(h,y) = \phi_{\mu_\lambda + \rho_H}(h)\left(\frac{1}{2}(1+y^2)\right)^{-a-\frac{n-1}{2}}$ continues analytically beyond $y = 0$.

Since

$$\langle \rho - w\lambda, Y_n\rangle = \begin{cases} \frac{n-1}{2} - a & \text{if } w = 1 \\ \frac{n-1}{2} + a & \text{if } w = -1 \end{cases}$$

formula (7.31) implies (7.24) in this case with $\varphi^\lambda_{\emptyset,1}(h,y) = 0$ in accordance with Theorem 7.6.3.

In the case $k = n-1$ the other function, $\psi_{-1,\lambda}$, is computed similarly.

Notice that the H-finiteness condition of Proposition 7.3.3 is $a + k - \frac{n-1}{2} \in \mathbb{Z}_+$ if $k < n-1$, and if $k = n-1$ all $\psi^H_{\pm 1,\lambda}$ are H-finite.

7.7.2 $G/H = SO_o(p,2)/SO_o(p-1,2)$

Let $G = SO_o(p,2)$ $(p \geq 2)$ and let H be the subgroup

$$H = \left\{\begin{pmatrix} 1 & 0 \\ 0 & A\end{pmatrix} \middle|\ A \in SO_o(p-1,2)\right\}$$

then G/H is one of the spaces of Example e in Section 7.1.

Let $X_{ij} = E_{ij} - E_{ji}$ and $Y_{ij} = E_{ij} + E_{ji}$, and define

$\mathfrak{h} = \mathrm{Span}\{X_{ij} \mid 1 \leq i,j \leq p \text{ or } i = p+1,\ j = p+2\}$,

$\mathfrak{a} = \mathrm{Span}\{Y_{p+1-j,p+j} \mid j = 1,2\}$, and

$\mathfrak{n} = \mathrm{Span}\{Y_{i,p+1} - X_{i,p},\ Y_{i,p+2} - X_{i,p-1},\ Z_1, Z_2 \mid 1 \leq i \leq p-2\}$, where

$$Z_1 = \begin{bmatrix} 0 & & 0 & & \\ & 0 & 1 & -1 & 0 \\ 0 & -1 & 0 & 0 & 1 \\ & -1 & 0 & 0 & 1 \\ & 0 & 1 & -1 & 0 \end{bmatrix} , \qquad Z_2 = \begin{bmatrix} 0 & & 0 & & \\ & 0 & 1 & -1 & 0 \\ 0 & -1 & 0 & 0 & -1 \\ & -1 & 0 & 0 & -1 \\ & 0 & -1 & 1 & 0 \end{bmatrix} .$$

Let $e_i \in \mathfrak{a}^*$ $(i = 1,2)$ be given by $e_i(Y_{p+1-j,p+j}) = \delta_{ij}$ $(j = 1,2)$. Then if $p > 2$ we have $\Sigma^+ = \{e_1; e_2; e_1 \pm e_2\}$ with multiplicities $p-2$ for e_j $(j = 1,2)$ and 1 for $e_1 \pm e_2$, whereas if $p = 2$ then $\Sigma^+ = \{e_1 \pm e_2\}$ with multiplicities 1.

We see that $\mathfrak{t} = \mathfrak{a} \cap \mathfrak{h}$ equals $\mathfrak{a}$ if $p > 2$ and $\mathbb{R} Y_{2,3}$ if $p = 2$. Thus $p = 2$ furnishes an example where the assumption of Section 7.6 does not hold. When $p \neq 3$ we have $W_\sigma = W_{K \cap H}(\mathfrak{a})$, when $p = 3$ there are two elements in $W_\sigma / W_{K \cap H}(\mathfrak{a})$.

Let $\mathfrak{b} = \mathbb{R} Y_{1,p+2}$ then $\mathfrak{b}$ is maximal abelian in $\mathfrak{p} \cap \mathfrak{q}$. We will only compute explicitly the restriction of ψ_λ to B. Because $K \cap H = SO(p-1) \times SO(2)$ and the $SO(p-1)$ factor commutes with $\mathfrak{b}$, the integral defining $\psi_\lambda(xK)$ for $x \in B$ reduces to an integration over $SO(2)$. For $t, \theta \in \mathbb{R}$ let

$$x(t) = \exp t\, Y_{1,p+2} = \begin{bmatrix} \cosh t & & & \sinh t \\ & 1 & & \\ & & \ddots & \\ & & & 1 \\ \sinh t & & & \cosh t \end{bmatrix} \quad \text{and}$$

$$k(\theta) = \exp \theta X_{p+1,p+2} = \begin{bmatrix} 1 & & & & \\ & \ddots & & & \\ & & 1 & & \\ & & & \cos\theta & \sin\theta \\ & & & -\sin\theta & \cos\theta \end{bmatrix}$$

then

$$\psi_\lambda(x(t)K) = \frac{1}{2\pi}\int_{-\pi}^{\pi} \exp\langle -\lambda-\rho, H(x(-t)k(\theta))\rangle \, d\theta \quad . \tag{7.32}$$

We compute $H(g)$ for $g \in G$ as follows. Define vectors $u_1, u_2 \in \mathbb{R}^{p+2}$ by

$$u_1 = \frac{1}{\sqrt{2}} \begin{bmatrix} 0 \\ \vdots \\ 0 \\ 0 \\ 1 \\ 1 \\ 0 \end{bmatrix} \quad \text{and} \quad u_2 = \frac{1}{\sqrt{2}} \begin{bmatrix} 0 \\ \vdots \\ 0 \\ 1 \\ 0 \\ 0 \\ 1 \end{bmatrix} .$$

In the natural representation of $so(p,2)$ on $\mathbb{R}^{p+2}$ we have $Z u_1 = 0$ for all $Z \in \mathfrak{n}$. Therefore $\langle e_1, H(g)\rangle = \log \|g u_1\|$. In the representation of $so(p,2)$ on $\mathbb{R}^{p+2} \wedge \mathbb{R}^{p+2}$ we have $Z(u_1 \wedge u_2) = 0$ for all $Z \in \mathfrak{n}$ and hence $\langle e_1 + e_2, H(g)\rangle = \log \|g(u_1 \wedge u_2)\|$. Recall that the norm on $\mathbb{R}^{p+2} \wedge \mathbb{R}^{p+2}$ is defined by

$$\|u \wedge v\|^2 = \|u\|^2 \|v\|^2 - \langle u,v\rangle^2 .$$

Letting $\lambda = a_1 e_1 + a_2 e_2$ $(a_1, a_2 \in \mathbb{C})$ we thus get

$$\langle \lambda, H(g)\rangle = \log \|g u_1\| (a_1 - a_2) + \log \|g(u_1 \wedge u_2)\| a_2$$

and hence

$$\langle \lambda + \rho, H(g)\rangle = \log \|g u_1\| (a_1 - a_2 + 1) + \log \|g(u_1 \wedge u_2)\| (a_2 + \tfrac{p}{2} - 1)$$

since $\rho = \frac{p}{2} e_1 + (\frac{p}{2} - 1) e_2$.

From (7.32) we then get

$$(7.33) \quad \frac{1}{2\pi} \int_{-\pi}^{\pi} \|x(-t)k(\theta)u_1\|^{-a_1 + a_2 - 1} \|x(-t)k(\theta)(u_1 \wedge u_2)\|^{-a_2 - \frac{p}{2} + 1} d\theta .$$

In case of $p = 3$ the other function $\psi_{w,\lambda}$ is computed similarly.

Assume first $p > 2$. By elementary computations

$$(7.34) \qquad \|x(-t)k(\theta)u_1\|^2 = 1 + \sinh^2 t \sin^2 \theta$$

$$(7.35) \qquad \|x(-t)k(\theta)u_1 \wedge u_2)\| = \cosh t$$

which inserted into (7.33) gives

$$\psi_\lambda(x(t)K) = \frac{1}{2\pi} (\cosh t)^{-a_2 - \frac{p}{2} + 1} \int_{-\pi}^{\pi} (1 + \sinh^2 t \sin^2 \theta)^{-\frac{1}{2}(a_1 - a_2 + 1)} d\theta .$$

By the substitution $u = \sin^2 \theta$ we get

$$\psi_\lambda(x(t)K) = \frac{1}{\pi}(\cosh t)^{-a_2-\frac{p}{2}+1}\int_0^1 u^{-\frac{1}{2}}(1-u)^{-\frac{1}{2}}(1+u\sinh^2 t)^{-\frac{1}{2}(a_1-a_2+1)}\,du$$

$$= (\cosh t)^{-a_2-\frac{p}{2}+1} F(\tfrac{1}{2}, \tfrac{1}{2}(a_1-a_2+1);1;-\sinh^2 t)$$

where F denotes the hypergeometric function (cf. Erdélyi et al. [a] p. 59 eq. (10)).

The asymptotic expansion of the hypergeometric function is well known. Putting $z = \sinh t$ we get from loc. cit. p. 63 eq. (17) that when $a_1-a_2 \notin 2\mathbf{Z}$:

$$\psi_\lambda(x(t)K) = \frac{1}{\sqrt{\pi}}(1+z^2)^{-\frac{1}{2}(a_2+\frac{p}{2}-1)}\left[z^{-1}\frac{\Gamma\left(\frac{a_1-a_2}{2}\right)}{\Gamma\left(\frac{a_1-a_2+1}{2}\right)} F(1,\tfrac{1}{2};1-\tfrac{1}{2}(a_1-a_2);-z^{-2})\right.$$

$$\left.+ z^{-(a_1-a_2+1)}\frac{\Gamma\left(\frac{a_2-a_1}{2}\right)}{\Gamma\left(\frac{a_2-a_1+1}{2}\right)} F(\tfrac{1}{2}(a_1-a_2+1),\tfrac{1}{2}(a_1-a_2+1),1+\tfrac{1}{2}(a_1-a_2);-z^{-2})\right].$$

With $t = -\log y$ we have asymptotically $z \sim \frac{1}{2}y^{-1}$ as $y \to 0$, whence

$$\psi_\lambda(b_y K) = \varphi_1(y)y^{\frac{p}{2}+a_2} + \varphi_2(y)y^{\frac{p}{2}+a_1} \tag{7.36}$$

where φ_j $(j = 1,2)$ are analytic at $y = 0$ with the values

$$\varphi_1(0) = \frac{1}{\sqrt{\pi}}\,2^{a_2+\frac{p}{2}}\frac{\Gamma\left(\frac{a_1-a_2}{2}\right)}{\Gamma\left(\frac{a_1-a_2+1}{2}\right)}, \quad \varphi_2(0) = \frac{1}{\sqrt{\pi}}\,2^{a_1+\frac{p}{2}}\frac{\Gamma\left(\frac{a_2-a_1}{2}\right)}{\Gamma\left(\frac{a_2-a_1+1}{2}\right)}.$$

Since $\langle \rho - w\psi, Y_{1,p}\rangle$ is either $\frac{p}{2} \pm a_1$ or $\frac{p}{2} \pm a_2$ we recover the asymptotic expansion given by Theorems 7.5.1 and 7.6.3.

It is interesting to notice the condition from Proposition 7.3.3 that ψ_λ is H-finite. Since $\mu_\lambda = (a_1 - \frac{p}{2}+1\ ,\ a_2 - \frac{p}{2}+2)$ we have $\mu_\lambda \in M^+$ if and only if

$$a_1 = \frac{p}{2} - 1 + j\ , \quad a_2 = a_1 - 1 - 2k$$

where j and k are integers, $j \geq 2k \geq 0$ if $p > 3$, and $j \geq k \geq 0$ if $p = 3$. Then $a_1 - a_2 = 2k+1$, so (7.36) holds. However, $\frac{1}{2}(a_2 - a_1 + 1) = -k$ and hence $\varphi_2(0) = 0$, so that we can improve (7.36) to

$$\psi_\lambda(b_y K) = \varphi_1(y) y^{\frac{p}{2} + a_2} .$$

Notice also that in this case the expression of $\psi_\lambda(x(t)K)$ as a hypergeometric function can be reduced to an expression involving a polynomial, in fact

$$F(\tfrac{1}{2}, k+1; 1; x) = \frac{1}{k!} \frac{d^k}{dx^k} (x^k (1-x)^{-1/2})$$

(cf. Erdélyi et al. [a] p. 102 eq. (21)).

Now we treat the case $p = 2$. There (7.33) still holds but instead of (7.34) we get

$$\| x(-t) k(\theta) (u_1 \wedge u_2) \| = \cosh t - \sinh t \cos \theta .$$

Inserted into (7.32) this gives

$$\psi_\lambda(x(t)K) = \frac{1}{2\pi} \int_{-\pi}^{\pi} (1 + \sinh^2 t \sin^2 \theta)^{-\frac{1}{2}(a_1 - a_2 + 1)} (\cosh t - \sinh t \cos \theta)^{-a_2} \, d\theta$$

$$= \frac{1}{2\pi} \int_{-\pi}^{\pi} (\cosh t + \sinh t \cos \theta)^{-\frac{1}{2}(a_1 - a_2 + 1)} (\cosh t - \sinh t \cos \theta)^{-\frac{1}{2}(a_1 + a_2 + 1)} \, d\theta.$$

Substituting $u = \frac{1}{2}(\cos \theta + 1)$ we get

$$\psi_\lambda(x(t)K)$$

$$= \frac{1}{\pi} e^{-a_2 t} \int_0^1 (1 - (1 - e^{2t}) u)^{-\frac{1}{2}(a_1 - a_2 + 1)} (1 - (1 - e^{-2t}) u)^{-\frac{1}{2}(a_1 + a_2 + 1)} u^{-\frac{1}{2}} (1-u)^{-\frac{1}{2}} \, du$$

$$= e^{-a_2 t} F_1(\tfrac{1}{2}, \tfrac{1}{2}(a_1 - a_2 + 1), \tfrac{1}{2}(a_1 + a_2 + 1), 1; 1 - e^{2t}, 1 - e^{-2t})$$

where F_1 is a hypergeometric function in two variables, cf. Erdélyi et al. [a] p. 231 eq. (5).

The asymptotic expansion of this function is found from loc. cit. p. 241 eqs. (10) and (11), which imply

$$\psi_\lambda(b_y K) = \varphi_1(y) y^{1+a_2} + \varphi_2(y) y^{1-a_2} + \varphi_3(y) y^{1+a_1} \tag{7.37}$$

where φ_j $(j=1,2,3)$ are analytic at $y=0$ with the values

$$\varphi_1(0) = \frac{\Gamma\left(\frac{a_1-a_2}{2}\right)}{\sqrt{\pi}\,\Gamma\left(\frac{a_1-a_2+1}{2}\right)} \;, \qquad \varphi_2(0) = \frac{\Gamma\left(\frac{a_1+a_2}{2}\right)}{\sqrt{\pi}\,\Gamma\left(\frac{a_1+a_2+1}{2}\right)}$$

$$\varphi_3(0) = \frac{\Gamma\left(\frac{-a_1+a_2}{2}\right)\Gamma\left(\frac{-a_1-a_2}{2}\right)}{\pi\,\Gamma(-a_1)} \;.$$

The formula (7.37) is valid whenever $\frac{1}{2}(a_1 \pm a_2)$ are not integers.

Again we recover the asymptotic expansion given by Theorem 7.5.1, with one term vanishing. Compared with (7.35) there is an extra term, y^{1-a_2}, which corresponds to the fact that the equal rank assumption of Theorem 7.6.3 fails.

With $p=2$ the condition which ensures that ψ_λ is H-finite is $a_1 \in \mathbb{Z}_+$. If this is the case, we see that $\varphi_3(0) = 0$ and the term involving y^{1+a_1} in (7.37) vanishes.

7.8 Notes and further results.

The first systematic study of nonRiemannian symmetric spaces was done by K. Nomizu [a], followed by M. Berger [a].

Proposition 7.1.1 occurs in Berger [a] p. 100. According to [a], it is also proved in Karpelevič [a] p. 19. Proposition 7.1.2 is due to G. D. Mostow [a], our proof is modelled after Loos [a] p. 161. The "Cartan decomposition" of Proposition 7.1.3 appears first in this form in Flensted-Jensen [a] and [b]. (See also Mostow [b] p. 262, Berger [a] p. 165 and Koh [a] for related statements, valid for arbitrary affine symmetric spaces, and Hoogenboom [a] for a similar result for compact symmetric spaces.) The remaining parts of Section 7.1 (Lemmas 7.1.4, 7.1.5 and Proposition 7.1.7) are due to T. Matsuki [a], partial results (Proposition 7.1.7(ii)) independently in Rossmann [b]. See also Aomoto [a] and Wolf [b], [c], for earlier results. The decomposition of G/P into H-orbits is generalized to the situation where P is an arbitrary parabolic subgroup in Matsuki [b]. In Oshima and Matsuki [a] G/H is decomposed into H-orbits, generalizing results of Kostant and Rallis [a] (See also van Dijk [a]).

Proposition 7.2.1 is due to W. Rossmann [b] (in Araki [a] a similar result is proved for σ-normal systems (cf. Warner [a]), but this can not be applied here). Lemma 7.2.3 is from Satake [a].

The functions $\psi_{w,\lambda}$ were introduced by M. Flensted-Jensen [c], where Lemma 7.3.1, Proposition 7.3.3 and Theorem 7.4.1 are proved. Theorem 7.3.4 is due to S. Helgason [c] (Results of this type also occur in Cartan [a] and Sugiura [a]). A generalization has been given by the author in [e]. Proposition 7.3.5 is new. For the case $H=K$ it says that the spherical functions satisfy $\phi_{s\nu}=\phi_{\nu}$, which was proved by Harish-Chandra ([c]I) with a different proof. Our proof is similar to Karpelevič [b] p. 158. The asymptotic expansions given in Theorems 7.5.1 and 7.6.3, and also Corollary 7.6.4 are due to T. Oshima (unpublished, [c]). Lemma 7.6.1 and Proposition 7.6.2 are from Flensted-Jensen [c]. In Flensted-Jensen [d] (see also [e]) a generalization of the functions $\psi_{w,\lambda} = \mathcal{P}T_{w,\lambda}$ is proposed, which is related to the non-closed H-orbits in G/P.

In addition to the Riemannian symmetric spaces, the semisimple symmetric spaces that have been most extensively studied are the hyperbolic spaces (i.e., $SO(p,q)/SO(p,q-1)$ and their complex and quaternion counterparts). See, e.g., Helgason [a], Gel'fand et al. [a], Wolf [a] and the references given in the Notes to Chapter 8. In Oshima and Sekiguchi [a] a quite extensive class of semisimple symmetric spaces is treated in the spirit of Chapter 5. In addition to the references given in the Notes to Chapters 5 and 8, other results on semisimple symmetric spaces can be found, for instance in Shapiro [a] and Hoogenboom [b]. For results on other symmetric spaces, see, e.g., Benoist [a], Cahen and Parker [a], and the references given there.

8. Construction of functions with integrable square

Let G/H be a semisimple symmetric space. Since H is reductive it follows from Helgason [n] Chapter 1, Theorem 1.9 that G/H has an invariant measure, unique up to scalars. Hence the Hilbert space $L^2(G/H)$ makes sense, and we can study the unitary representation

$$(\pi(g)f)(xH) = f(g^{-1}xH)$$

$(g, x \in G)$ of G on this space. It is the purpose of L^2-harmonic analysis on G/H to give an explicit decomposition (in general as a direct integral) of this representation into irreducibles. So far this program has not been accomplished in general (although the answer is known in several specific cases, notably those of $L^2(G/K)$ and $L^2(G \times G/d(g)) \cong L^2(G)$, by the work of Harish-Chandra - see the notes at the end of this chapter).

In this book we content ourselves with a less ambitious task, namely that of pointing out some representations that enter discretely (see below) into $L^2(G/H)$. Even though restricting ourselves to the discrete part of $L^2(G/H)$, we shall not give a complete description, but only construct the "simplest" part of the discrete series (in some special cases, though, e.g., $G \times G/d(G)$, the construction gives the complete discrete series).

By definition, the discrete series for G/H consists of those (equivalence classes of) unitary irreducible representations of G, which are realized as subrepresentations of π on closed subspaces of $L^2(G/H)$. (This means that in the decomposition of $L^2(G/H)$ as a direct integral, the discrete series enters as a sum).

Using the fundamental functions defined in the previous chapter and a certain duality given in Section 8.2, we will construct a family of square integrable functions on the symmetric space G/H, provided a certain rank condition holds (Theorem 8.3.1). We then show that the representations of G generated by these functions belong to the discrete series.

8.1 The invariant measure on G/H

At each point $x = gH$ of the semisimple symmetric space G/H we can identify the tangent space $T_x(G/H)$ with $\mathfrak{q}$ via the differential of the map $\mathfrak{q} \ni X \longrightarrow g \exp X H \in G/H$. On $\mathfrak{q}$ Killing form gives a bilinear form which is nondegenerate, and it hence follows that the space G/H can be given a structure of a pseudo-Riemannian manifold. Associated to the G-invariant pseudo-Riemannian metric is a G-invariant measure on G/H, which we call the normalized invariant measure.

Let $\mathfrak{b} \subset \mathfrak{p} \cap \mathfrak{q}$ be a maximal abelian subspace, and let $M_b = Z_{K \cap H}(\mathfrak{b})$ denote the centralizer of $\mathfrak{b}$ in $K \cap H$. Then we have from Proposition 7.1.3 that the map

$$\Phi : K/M_b \times B_o^+ \ni (kM_b, b) \longrightarrow kbH \in G/H$$

is a diffeomorphism onto an open dense subset of G/H. This map can be interpreted as "polar coordinates" on G/H. It is the purpose of this section to relate the invariant measure on G/H to Haar measure on K and B via Φ.

Example Let $G = SO_o(1,2)$ and $H = SO_o(1,1)$. Then

$$G/H \cong \{x \in \mathbb{R}^3 \mid -x_1^2 + x_2^2 + x_3^2 = 1\}$$

and we have $K = SO(2)$, $K \cap H = \{e\}$, $B = B_o^+ \cong \mathbb{R}$, and

$$\Phi(e^{i\alpha}, t) = (\sinh t\ ,\ \sin\alpha \cosh t\ ,\ \cos\alpha \cosh t)$$

for $e^{i\alpha} \in SO(2)$ and $t \in \mathbb{R}$. On $\mathbb{R}^3$ the measure $dx_1dx_2dx_3$ is invariant for $SO_o(1,2)$. Writing $\xi = -x_1^2 + x_2^2 + x_3^2$ we have

$$dx_1dx_2dx_3 = \frac{1}{2} x_3^{-1} dx_1dx_2d\xi \ .$$

Therefore $x_3^{-1}dx_1dx_2$ is invariant on

$$\{x \in \mathbb{R}^3 \mid -x_1^2 + x_2^2 + x_3^2 = 1\ ,\ x_3 \neq 0\} \ .$$

From this it follows that in terms of the "polar coordinates" α and t, the measure

$$\cosh t \ d\alpha dt$$

is invariant on G/H. □

For each $\beta \in \Sigma_b = \Sigma(\mathfrak{b}, \mathfrak{g})$ the root space $\mathfrak{g}^\beta$ is invariant under $\sigma\theta$, and hence decomposes as follows:

$$\mathfrak{g}^\beta = (\mathfrak{g}^\beta \cap \mathfrak{g}_o) \oplus (\mathfrak{g}^\beta \cap (\mathfrak{k} \cap \mathfrak{q} + \mathfrak{p} \cap \mathfrak{h})) .$$

We denote $p_\beta = \dim \mathfrak{g}^\beta \cap \mathfrak{g}_o$ and $q_\beta = \dim(\mathfrak{g}^\beta \cap (\mathfrak{k} \cap \mathfrak{q} + \mathfrak{p} \cap \mathfrak{h}))$. Then p_β, resp. $p_\beta + q_\beta$, is the multiplicity of β in $\Sigma_o = \Sigma(\mathfrak{b}, \mathfrak{g}_o)$, resp. Σ_b.

Let $\delta(Y) = \left| \prod_{\beta \in \Sigma_b^+} (\sinh \beta(Y))^{p_\beta} (\cosh \beta(Y))^{q_\beta} \right|$ for $Y \in \mathfrak{b}$.

(Notice that $\delta(Y)$ is independent of the actual choice of positive system Σ_b^+ for Σ_b).

<u>Theorem 8.1.1</u> <u>The normalized invariant measure on</u> G/H <u>is given by</u>

$$\int_{G/H} f(gH)\,dgH = \int_K \int_{\mathfrak{b}_o^+} f(k \exp Y H)\,\delta(Y)\,dY\,dk$$

<u>for</u> $f \in C_c(G/H)$, <u>where</u> dY <u>denotes Lebesgue measure on</u> $\mathfrak{b}$ <u>and</u> dk <u>Haar measure on</u> K, <u>normalized by the Killing form.</u>

<u>Proof</u>: We have to prove that $\delta(Y)$ is the Jacobian of Φ at $(kM_b, \exp Y)$ for all $k \in K$, $Y \in \mathfrak{b}$, with respect to the Killing form.

Let $\mathfrak{k}' \subset \mathfrak{k}$ denote the orthocomplement in $\mathfrak{k}$ of the centralizer $\mathfrak{m}_b = \mathfrak{k} \cap \mathfrak{h}^{\mathfrak{b}}$ of $\mathfrak{b}$ in $\mathfrak{k} \cap \mathfrak{h}$. We identify the tangent spaces of K/M_b and B_o^+ at kM_b and b, respectively, with $\mathfrak{k}'$ and $\mathfrak{b}$ via the differentials of $\mathfrak{k}' \ni X \longrightarrow k \exp X M_b \in K/M_b$ and $\mathfrak{b} \ni Y \longrightarrow b \exp Y \in B$. Since

$$\Phi(k \exp X M_b, b \exp Y) = kb \exp(\mathrm{Ad}\, b^{-1} X) \exp Y H$$

it follows that $d\Phi_{(kM_b, b)}(X,Y)$ is given by the projection to $\mathfrak{q}$ along $\mathfrak{h}$ of the vector $\mathrm{Ad}\, b^{-1} X + Y$ in $\mathfrak{g}$. We will now determine this explicitly in terms of bases for $\mathfrak{k}'$, $\mathfrak{b}$, and $\mathfrak{q}$.

For simplicity of notation we use the convention that $\Sigma_b^{+\prime}$ consists of the roots from Σ_b^+, each repeated according to its multiplicity $p_\beta + q_\beta$. For each $\beta \in \Sigma_b^{+\prime}$ we pick $X_\beta \in \mathfrak{g}^\beta$ such that $\sigma\theta X_\beta = \pm X_\beta$ and such that the various X_β corresponding to the same element of Σ_b^+ form a basis for the root space. Let $X_{-\beta} = \theta X_\beta$, $X'_\beta = X_\beta + X_{-\beta}$, and $X''_\beta = X_\beta - X_{-\beta}$. If $\sigma\theta X_\beta = X_\beta$ then $X'_\beta \in \mathfrak{h} \cap \mathfrak{h}$ and $X''_\beta \in \mathfrak{p} \cap \mathfrak{q}$, whereas if $\sigma\theta X_\beta = -X_\beta$ then $X'_\beta \in \mathfrak{h} \cap \mathfrak{q}$ and $X''_\beta \in \mathfrak{p} \cap \mathfrak{h}$.

The elements X'_β $(\beta \in \Sigma_b^{+\prime})$ together with a basis for the centralizer $\mathfrak{h}'^{\mathfrak{b}}$ form a basis for $\mathfrak{h}'$. On the other hand, as a basis for $\mathfrak{q}$ we can use the elements X'_β $(\beta \in \Sigma_b^{+\prime}, \sigma\theta X_\beta = -X_\beta)$ and X''_β $(\beta \in \Sigma_b^{+\prime}, \sigma\theta X_\beta = X_\beta)$ together with the same basis for $\mathfrak{h}'^{\mathfrak{b}}$ as before, and some basis for $\mathfrak{b}$.

With $b = \exp Y_o$, $Y_o \in \mathfrak{b}$, it is easily seen that

$$\mathrm{Ad}\, b^{-1} X'_\beta = \cosh\beta(Y_o)\, X'_\beta - \sinh\beta(Y_o)\, X''_\beta .$$

Hence it follows that:

<u>Lemma 8.1.2</u> <u>The differential</u> $d\Phi$ <u>of</u> Φ <u>is given by</u>

$$d\Phi_{(kM_b,b)}(X'_\beta) = \begin{cases} \cosh\beta(Y_o)\, X'_\beta & \underline{\text{if}}\ \ \sigma\theta X_\beta = -X_\beta \\ -\sinh\beta(Y_o)\, X''_\beta & \underline{\text{if}}\ \ \sigma\theta X_\beta = X_\beta \end{cases}$$

<u>and</u> $d\Phi_{(kM_b,b)}(Z) = Z$ <u>if</u> $Z \in \mathfrak{h}'^{\mathfrak{b}}$ <u>or</u> $Z \in \mathfrak{b}$.

Since the bases for $\mathfrak{h}'$, $\mathfrak{b}$ and $\mathfrak{q}$ described above can be chosen orthonormally with respect to Killing form, it follows from this lemma that $\delta(Y_o)$ is the Jacobian of Φ at $(kM_b, \exp Y_o)$, which concludes the proof of Theorem 8.1.1. □

8.2 <u>An important duality</u>

As is well known, there is a certain duality due to E. Cartan between Riemannian symmetric space of respectively the noncompact and the compact type. If $\mathfrak{g}$ is a semisimple noncompact Lie algebra with Cartan decomposition $\mathfrak{g} = \mathfrak{h} \oplus \mathfrak{p}$, then the compact form $\mathfrak{u} = \mathfrak{h} \oplus \sqrt{-1}\,\mathfrak{p}$ in $\mathfrak{g}_c$ is "dual" to $\mathfrak{g}$, and vice versa (Helgason [j] Chapter V). In this section we present what might be viewed as the generalization of this duality to semisimple symmetric spaces.

Let $\mathfrak{g}$ be a real semisimple Lie algebra, let σ be an involution of $\mathfrak{g}$, let $\mathfrak{k}$ be a maximal compact subalgebra invariant under σ, and let θ be the corresponding Cartan involution. As before, we have the decompositions $\mathfrak{g} = \mathfrak{k} \oplus \mathfrak{p}$ and $\mathfrak{g} = \mathfrak{h} \oplus \mathfrak{q}$. We now define subalgebras $\mathfrak{g}^o$, $\mathfrak{h}^o$, and $\mathfrak{k}^o$ of $\mathfrak{g}_c$ by:

$$\mathfrak{g}^o = \mathfrak{k} \cap \mathfrak{h} + \mathfrak{p} \cap \mathfrak{q} + \sqrt{-1}\,(\,\mathfrak{k} \cap \mathfrak{q} + \mathfrak{p} \cap \mathfrak{h}\,)\,,$$

$$\mathfrak{h}^o = \mathfrak{k} \cap \mathfrak{h} + \sqrt{-1}\,(\,\mathfrak{p} \cap \mathfrak{h}\,)\,, \quad \text{and}$$

$$\mathfrak{k}^o = \mathfrak{k} \cap \mathfrak{h} + \sqrt{-1}\,(\,\mathfrak{k} \cap \mathfrak{q}\,)\,.$$

Then the triple consisting of $\mathfrak{g}^o$, $\mathfrak{h}^o$, and $\mathfrak{k}^o$ is called dual to the triple of $\mathfrak{g}$, $\mathfrak{k}$, and $\mathfrak{h}$. Notice that $\mathfrak{k}^o$ is maximally compact in $\mathfrak{g}^o$, and that $\mathfrak{h}^o$ consists of the fixed points in $\mathfrak{g}^o$ for the involution derived by restricting to $\mathfrak{g}^o$ the complex linear extension of θ to $\mathfrak{g}_c$. It follows easily that the duality is symmetric in the sense that $(\mathfrak{g}, \mathfrak{k}, \mathfrak{h})$ is dual to $(\mathfrak{g}^o, \mathfrak{h}^o, \mathfrak{k}^o)$.

Notice that if $\mathfrak{g} = \mathfrak{h}$ and $\mathfrak{g}$ is noncompact, then $\mathfrak{g}^o = \mathfrak{h}^o = \mathfrak{k} + \sqrt{-1}\,\mathfrak{p} = \mathfrak{u}$, and thus Cartan's duality is a special case of this duality. If $\mathfrak{k} = \mathfrak{h}$ then $\mathfrak{g}^o = \mathfrak{g}$ and $\mathfrak{h}^o = \mathfrak{k}^o = \mathfrak{k}$.

Let G_c be a connected Lie group with Lie algebra $\mathfrak{g}_c$, and let G, G^o, H etc. be the real analytic subgroups corresponding to $\mathfrak{g}, \mathfrak{g}^o, \mathfrak{h}$ etc. Then we also say that (G,K,H) and (G^o,H^o,K^o) are dual to each other. Notice that G^o/H^o is a Riemannian symmetric space, that $K^o \cap H^o = K \cap H$, and that G_o is the identity component of $G \cap G^o$. The purpose of introducing this duality is to move analysis from the pseudo-Riemannian space G/H to the Riemannian space G^o/H^o, where the results of the preceding chapters can be used.

Let $\hat{K}$ and $\hat{K}^o$ denote the sets of equivalence classes of irreducible finite dimensional representations of K and K^o, respectively, and for $\delta \in \hat{K}$ let $C_\delta^\infty(G/H)$ denote the linear span in $C^\infty(G/H)$ of all functions K-finite of type δ. Let $C_K^\infty(G/H) = \bigoplus_{\delta \in \hat{K}} C_\delta^\infty(G/H)$ be the space of all K-finite C^∞ functions on G/H (the sum is an algebraic direct sum). Define $C_{\delta^o}^\infty(G^o/H^o)$ for $\delta^o \in \hat{K}^o$ and $C_{K^o}^\infty(G^o/H^o)$ similarly.

The space $C_K^\infty(G/H)$ is invariant under $\mathfrak{g}$ acting from the left, for if $f \in C_K^\infty(G/H)$ and $S \in \mathfrak{g}$ then

(8.1) $$(Sf)(kx) = (\mathrm{Ad}k^{-1}S)f^k(x)$$

for $k \in K$ and $x \in G$ where f^k is given by $f^k(x) = f(kx)$, and hence Sf is K-finite by the finite dimensionality of $\mathfrak{g}$. By complexification $C_K^\infty(G/H)$ is a $\mathfrak{g}_c$-module. Similarly, $C_{K^o}^\infty(G^o/H^o)$ is $\mathfrak{g}^o$-invariant, and hence a $\mathfrak{g}_c$-module by complexification.

Denote by $\mathbb{D}(G/H)$ the algebra of differential operators on G/H invariant for the action of G. Let $U(\mathfrak{g})^H$ denote the centralizer of H in $U(\mathfrak{g})$, then right action gives a canonical homomorphism of $U(\mathfrak{g})^H$ onto $\mathbb{D}(G/H)$. The kernel is $U(\mathfrak{g})^H \cap U(\mathfrak{g})\mathfrak{h}_c$ which is also the kernel of the canonical homomorphism of $U(\mathfrak{g})^{H^o}$ onto $\mathbb{D}(G^o/H^o)$. We therefore have an isomorphism $D \to D^o$ of $\mathbb{D}(G/H)$ with $\mathbb{D}(G^o/H^o)$. In particular $\mathbb{D}(G/H)$ is commutative.

Obviously, each space $C_\delta^\infty(G/H)$ is invariant under $\mathbb{D}(G/H)$, and similarly $C_{\delta^o}^\infty(G^o/H^o)$ is invariant under $\mathbb{D}(G^o/H^o)$.

Let K_c be the analytic subgroup of G_c with Lie algebra $\mathfrak{k}_c$, and let $\hat{K}_c$ denote the set of equivalence classes of irreducible holomorphic finite dimensional representations of K_c. By restriction we have injections $\hat{K}_c \to \hat{K}$ and $\hat{K}_c \to \hat{K}^o$, the former being surjective, the latter, however, not surjective in general (unless K_c is exchanged with some covering group). Let $\hat{K}^o(K_c)$ denote the image of $\hat{K}_c \to \hat{K}^o$. For $\delta \in \hat{K}$ let $\delta^o \in \hat{K}^o$ be the restriction to K^o of the extension to K_c of δ, then $\delta \to \delta^o$ is a bijection $\hat{K} \to \hat{K}^o(K_c)$.

<u>Theorem 8.2.1</u> <u>The subspace</u> $\bigoplus_{\delta^o \in \hat{K}^o(K_c)} C_{\delta^o}^\infty(G^o/H^o)$ <u>of</u> $C_{K^o}^\infty(G^o/H^o)$ <u>is</u> $\mathfrak{g}_c$<u>-invariant, and isomorphic as a</u> $\mathfrak{g}_c$<u>-module to</u> $C_K^\infty(G/H)$ <u>with an isomorphism</u> $f^o \leftrightarrow f$ <u>satisfying</u> $f(y) = f^o(y)$ <u>for</u> $y \in G_o$ <u>and</u> $f^o \in C_{\delta^o}^\infty(G^o/H^o)$ <u>when</u> $f \in C_\delta^\infty(G/H)$. <u>Moreover</u> $(Df)^o = D^o f^o$ <u>for</u> $D \in \mathbb{D}(G/H)$.

<u>Proof:</u> Let $f \in C_K^\infty(G/H)$ and let $E \subset C_K^\infty(G/H)$ denote the finite dimensional complex linear span of the K-translates of f. Let π be the corresponding representation of K, then π extends holomorphically to a representation π_c of K_c on E.

For each function $\varphi \in E$ we define $\varphi^o \in C^\infty(G^o/H^o)$ by

$$(8.2) \qquad \varphi^o(x) = (\pi_c(\exp X)^{-1}\varphi)(\exp Y H)$$

for $x = \exp X \exp Y H^o$, $X \in \sqrt{-1}(\mathfrak{h} \cap \mathfrak{q})$ and $Y \in \mathfrak{p} \cap \mathfrak{q}$, using Proposition 7.1.2. Obviously $\varphi \to \varphi^o$ is a linear map and $\varphi^o(y) = \varphi(y)$ for $y \in G_o = \exp(\mathfrak{p} \cap \mathfrak{q})K \cap H$. We claim that

$$(8.3) \qquad (\pi_c(k^o)\varphi)^o(x) = \varphi^o(k^{o-1}x)$$

for $\varphi \in E$, $k^o \in K^o$ and $x \in G^o/H^o$. Let X and Y be as above and let $k^{o-1}\exp X = \exp X' \ell$ with $X' \in \sqrt{-1}(\mathfrak{h} \cap \mathfrak{q})$, $\ell \in K \cap H$. Then $k^{o-1}x = \exp X' \exp(\mathrm{Ad}\,\ell\, Y)H^o$ and hence

$$\begin{aligned}\varphi^o(k^{o-1}x) &= (\pi_c(\exp X')^{-1}\varphi)(\exp(\mathrm{Ad}\,\ell\, Y)H)\\ &= (\pi_c(\exp X')^{-1}\varphi)(\ell \exp Y H)\\ &= (\pi_c(k^{o-1}\exp X)^{-1}\varphi)(\exp Y H)\\ &= (\pi_c(k^o)\varphi)^o(x)\end{aligned}$$

as claimed. Therefore $\varphi \to \varphi^o$ is a K^o-map. If $\delta \in \hat{K}$ and $f \in C^\infty_\delta(G/H)$ we thus have $f^o \in C^\infty_{\delta^o}(G^o/H^o)$.

Conversely let $g \in C^\infty_{\delta^o}(G^o/H^o)$ $(\delta^o \in \hat{K}^o(K_c))$, and let E^o denote the space spanned by the K^o-translates of g. Let π^o be the corresponding representation of K^o, then π^o extends holomorphically to a representation π^o_c of K_c on E^o, since $\delta^o \in \hat{K}^o(K_c)$. We define ${}^o g \in C^\infty(G)$ by

$$(8.4) \qquad {}^o g(x) = (\pi^o_c(k)^{-1}g)(\exp Y)$$

for $x = k \exp Y \exp Z \in G$, $Y \in \mathfrak{p} \cap \mathfrak{q}$ and $Z \in \mathfrak{p} \cap \mathfrak{h}$, cf. Proposition 7.1.2. The proof that ${}^o g \in C^\infty(G/H)$ is similar to the proof of (8.3) above, and from (8.4) it is obvious that then ${}^o g \in C^\infty_\delta(G/H)$. It is now easily seen that ${}^o(f^o) = f$ and $({}^o g)^o = g$.

It remains to be seen that $f \to f^o$ is a $\mathfrak{g}_c \times U(\mathfrak{g})^H$ map. We need the following lemma.

<u>Lemma 8.2.2</u> <u>For</u> $y \in G_o$ <u>we have</u>

$$\mathfrak{g} = \mathfrak{h} \oplus \mathfrak{p} \cap \mathfrak{q} \oplus \mathrm{Ad}\,y(\mathfrak{p} \cap \mathfrak{h}) = \mathfrak{h} \oplus \mathfrak{p} \cap \mathfrak{q} \oplus \mathrm{Ad}\,y(\mathfrak{h} \cap \mathfrak{q}).$$

Proof: The second equality follows from the first by applying it to $\mathfrak{g}^o$. Since $\mathfrak{k}\cap\mathfrak{q} + \mathfrak{p}\cap\mathfrak{h}$ is invariant under ad $\mathfrak{g}_o$, we have that $\mathrm{Ad}\,y(\mathfrak{p}\cap\mathfrak{h}) \subset \mathfrak{k}\cap\mathfrak{q} + \mathfrak{p}\cap\mathfrak{h}$. By reasons of dimension we only have to prove that $\mathrm{Ad}\,y(\mathfrak{p}\cap\mathfrak{h}) \cap \mathfrak{k}\cap\mathfrak{q} = 0$. But $\mathrm{Ad}\,y(\mathfrak{p}) \cap \mathfrak{k} = 0$ for any $y \in G$ since $\mathrm{Ad}\,y$ preserves the Killing form. Thus the first equality holds. □

We will now show that $(Sf)^o(x) = Sf^o(x)$ for $S \in \mathfrak{g}_c$ and $x \in G^o/H^o$. Let $x = \exp X \exp Y H^o$ with $X \in \sqrt{-1}\,\mathfrak{k}\cap\mathfrak{q}$ and $Y \in \mathfrak{p}\cap\mathfrak{q}$, then it easily follows from (8.1) and (8.2) that

$$(Sf)^o(x) = [(\mathrm{Ad}(\exp X)^{-1}S)(\pi_c(\exp X)^{-1}f)](\exp Y) .$$

Let $S' = \mathrm{Ad}(\exp X)^{-1}S$ and $f' = \pi_c(\exp X)^{-1}f$, then we thus have

$$(8.5) \qquad (Sf)^o(x) = (S'f')(\exp Y) .$$

On the other hand, it easily follows from (8.3) that

$$(8.6) \qquad Sf^o(x) = S'f'^o(\exp Y) .$$

From (8.5) and (8.6) we see that we may assume $X = 0$.

Let $y = \exp Y$, $Y \in \mathfrak{p}\cap\mathfrak{q}$. We want to prove

$$(8.7) \qquad Sf(y) = Sf^o(y)$$

for all $S \in \mathfrak{g}_c$. If $S \in \mathfrak{k}_c$ this follows from (8.3). If $S \in \mathfrak{p}\cap\mathfrak{q}$ this is obvious since $f = f^o$ on G_o. Finally, if $S \in \mathrm{Ad}\,y(\mathfrak{p}\cap\mathfrak{h})_c$ we have $Sf(y) = Sf^o(y) = 0$ by the right invariance under H and H^o. By Lemma 8.2.2, (8.7) is proved and so $f \longrightarrow f^o$ is a $\mathfrak{g}_c$-map.

Let $D \in \mathbb{D}(G/H)$, we want to prove that $(Df)^o(x) = D^of^o(x)$ for $x \in G^o/H^o$. By (8.3) we may assume $x = \exp Y H^o$ where $Y \in \mathfrak{p}\cap\mathfrak{q}$. Then it suffices to show (8.7), this time with S acting from the right. This follows from Lemma 8.2.2 by an argument similar to that above. This completes the proof of Theorem 8.2.1. □

Remark 8.2.3 It follows from the preceding proof that for each $x \in G_o$ the function $k^o \longrightarrow f^o(k^ox)$ on K^o is the analytic continuation of the function $k \longrightarrow f(kx)$ on K, and vice versa.

Let $G^{(\sim)}$ be the simply connected covering group of G, and let $K^{(\sim)}$ and $H^{(\sim)}$ be the analytic subgroups corresponding to $\mathfrak{k}$ and $\mathfrak{h}$. Let $\eta_o : G^{(\sim)} \longrightarrow G$ be the covering map and let $Z_o = \eta_o^{-1}(e) \cap H^{(\sim)}$. Define $G^{\sim} = G^{(\sim)}/Z_o$ and $K^{\sim} = K^{(\sim)}/Z_o$. We see that we can identify H, $H \cap K$ and G_o with the corresponding subgroups of $G^{\sim}$. Notice that $G^{\sim}/H \simeq G^{(\sim)}/H^{(\sim)}$ is the simply connected covering space of G/H.

Let $K_c^{(\sim)}$ be the simply connected covering group of K_c. Since K is maximal compact in K_c we can consider $K^{(\sim)}$ as a subgroup of $K_c^{(\sim)}$ and hence define $K_c^{\sim} = K_c^{(\sim)}/Z_o$. We can then identify K^o and $K \cap H$ with the corresponding analytic subgroups of $K_c^{\sim}$.

The bijective correspondence $\delta \longleftrightarrow \delta^0$ between $\hat{K}$ and $\hat{K}^o(K_c)$ can now be extended to a bijective correspondence from the set $\hat{K}^{\sim}$ of equivalence classes of irreducible finite dimensional representations of $K^{\sim}$, to $\hat{K}^o$.

Proceeding exactly as in Theorem 8.2.1 we get

<u>Theorem 8.2.4</u> <u>There is an isomorphism</u> $f \to f^o$ <u>between the spaces</u> $C^{\infty}_{K^{\sim}}(G^{\sim}/H)$ <u>and</u> $C^{\infty}_{K^o}(G^o/H^o)$, <u>such that</u> $f(y) = f^o(y)$ <u>for</u> $y \in G_o$ <u>and</u> $f^o \in C^{\infty}_{\delta^o}(G^o/H^o)$ <u>for</u> $f \in C^{\infty}_{\delta}(G^{\sim}/H)$ <u>and</u> $\delta \in \hat{K}^{\sim}$. <u>Moreover</u> $f \to f^o$ <u>is a</u> $\mathfrak{g}_c$<u>-map and</u> $(Df)^o = D^o f^o$ <u>for</u> $D \in \mathbb{D}(G^{\sim}/H) \cong \mathbb{D}(G/H)$.

8.3 <u>Discrete series</u>

Let G be a semisimple connected noncompact Lie group with finite center, and let G/H be a symmetric space. We are now in position to construct discrete representations in $L^2(G/H)$.

A subspace $\mathfrak{a} \subset \mathfrak{q}$ is called a <u>θ-stable Cartan subspace</u> if $\mathfrak{a}$ is maximal abelian in $\mathfrak{q}$ and $\theta\mathfrak{a} = \mathfrak{a}$ (recall that θ is a Cartan involution commutative with the given involution σ). A subspace $\mathfrak{a} \subset \mathfrak{q}$ is called a <u>Cartan subspace</u> if it is maximal abelian and consists of semisimple elements. It is a fact, which we do not need here, that every Cartan subspace is conjugate by H to a θ-stable Cartan subspace (Oshima and Matsuki [a] p. 406, Remark).

Let $\mathfrak{a}$ be a θ-stable Cartan subspace of $\mathfrak{q}$, and let $\mathfrak{a}^o = \sqrt{-1}\ \mathfrak{a} \cap \mathfrak{k} + \mathfrak{a} \cap \mathfrak{p}$, then $\mathfrak{a}^o$ is a maximal abelian subspace of $\mathfrak{q}^o = \sqrt{-1}\, \mathfrak{q} \cap \mathfrak{h} + \mathfrak{q} \cap \mathfrak{p}$. We define the rank of G/H as the dimension of $\mathfrak{a}$. In case where G/H is Riemannian this definition coincides with the usual one. We have

$$\operatorname{rank} G/H = \operatorname{rank} G^o/H^o \quad .$$

In particular the rank does not depend on the choice of $\mathfrak{a}$.

We say that $\mathfrak{a}$ is fundamental if $\mathfrak{t} = \mathfrak{a} \cap \mathfrak{h}$ is a maximal abelian subspace of $\mathfrak{q} \cap \mathfrak{h}$, that is, if $\mathfrak{a}^o$ is $\mathfrak{h}^o$-maximal. By Lemma 7.1.5 all fundamental Cartan subspaces are mutually conjugate by $K \cap H$. Let $\mathfrak{a}$ be one such fundamental Cartan subspace, let $\Sigma = \Sigma(\mathfrak{a}, \mathfrak{g}_c)$ denote its root system and choose a positive system Σ^+ which is $\mathfrak{h}^o$-compatible, that is, $\alpha \in \Sigma^+$ and $\alpha|_{\mathfrak{t}} \neq 0$ implies that $\theta\alpha \in \Sigma^+$. Let W denote the Weyl group of Σ .

In order to apply the duality we assume first that G is linear, that is, $G \subset G_c$ where G_c has Lie algebra $\mathfrak{g}_c$. Let G^o/H^o denote the dual Riemannian symmetric space with $G^o \subset G_c$.

For each $\lambda \in \mathfrak{a}_c^*$ and $w \in W_\sigma = \{w \in W \mid w(\mathfrak{t}) = \mathfrak{t}\}$ we have defined in Section 7.3 the function $\psi^o_{w,\lambda} = \psi^{H^o}_{w,\lambda} \in C^\infty(G^o/H^o)$ by

$$\psi^o_{w,\lambda}(xH^o) = \int_{K \cap H} \exp\langle -\lambda-\rho, H(x^{-1}kw)\rangle \, dk$$

for $x \in G^o$. By Proposition 7.3.3, if $\mu_{w,\lambda} = w(\lambda+\rho)|_{\mathfrak{t}} - 2\rho_c \in M^+$, that is, if

$$(8.8) \qquad \frac{\langle \mu_{w,\lambda}, \beta\rangle}{\langle \beta, \beta\rangle} \in \mathbb{Z}_+ \quad \text{for all } \beta \in \Sigma_c^+$$

where $\Sigma_c = \Sigma(\mathfrak{t}_c, \mathfrak{h}_c)$, $\Sigma_c^+ = \{\beta \in \Sigma_c \mid \beta = \alpha|_{\mathfrak{t}_c}, \alpha \in \Sigma^+\}$ and $\rho_c = \frac{1}{2} \sum_{\beta \in \Sigma_c^+} (\dim \mathfrak{h}_c^\beta)\beta$, then $\psi^o_{w,\lambda}$ is K^o-finite of the irreducible K^o-type of highest weight $\mu_{w,\lambda}$ on $\sqrt{-1}\,\mathfrak{t}$. Moreover, if in addition we assume that

$$(8.9) \qquad X \in \mathfrak{t} \text{ and } \exp X = e \Longrightarrow \mu_{w,\lambda}(X) \in 2\pi\sqrt{-1}\,\mathbb{Z}$$

then it is easily seen that this K^o-type belongs to $\hat{K}^o(K_c)$. By Theorem 8.2.1, the fundamental function $\psi^o_{w,\lambda}$ then corresponds to a function $\psi_{w,\lambda} \in C_K^\infty(G/H)$.

Definition The functions $\psi_{w,\lambda} \in C_K^\infty(G/H)$ defined for each $w \in W_\sigma$ and $\lambda \in \mathfrak{a}_c^*$ satisfying (8.8) and (8.9) are called the (fundamental) Flensted-Jensen functions on G/H.

By Theorem 8.2.1 the K-type generated by $\psi_{w,\lambda}$ is the representation, spherical with respect to $K \cap H$, of highest weight $\mu_{w,\lambda}^\vee$ on $\mathfrak{t}$. Moreover, since $\psi_{w,\lambda}^o \in \mathcal{A}(G^o/H^o; \mathcal{M}_\lambda)$ and because of Theorem 7.4.1, $\psi_{w,\lambda}$ is a joint eigenfunction for all $D \in \mathbb{D}(G/H)$ (from the right) and $u \in U(\mathfrak{g})^K$ (from the left) with eigenvalues, respectively, $\chi_\lambda(D^o)$ and $\xi(u)(-\lambda-\rho)$, where $\xi(u) \in U(\mathfrak{a})$ is determined by $u - \xi(u) \in (\mathfrak{k} \cap \mathfrak{h})_c U(\mathfrak{g}) + U(\mathfrak{g})(\mathfrak{h}^{\mathfrak{a}} + \mathfrak{n})_c$.

Theorem 8.3.1 If rank G/H = rank $K/K \cap H$ and $\lambda \in \mathfrak{a}_+^*$, that is, $\langle \lambda, \alpha \rangle > 0$ for all $\alpha \in \Sigma^+$, then when (8.8) and (8.9) hold the Flensted-Jensen functions $\psi_{w,\lambda}$ $(w \in W_\sigma)$ are square integrable.

Proof: Exchanging Σ^+ by $w\Sigma^+$ and λ by $w\lambda$ we may assume $w = e$. Let $\varphi_{\mu_\lambda}(k^o) = \exp\langle \mu_\lambda, H(k^o)\rangle$ for $k^o \in K^o$, and let φ_{μ_λ} also denote the analytic continuation of this matrix coefficient of δ_{μ_λ} to K_c. Then by Lemma 7.3.1 we have

$$\psi_\lambda^o(k^o x H^o) = \int_{K \cap H} \varphi_{\mu_\lambda}(k^o \ell) \exp\langle -\lambda-\rho, H(x^{-1}\ell)\rangle \, d\ell$$

for $x \in G^o$. For $x \in G_o$ it then follows from Remark 8.2.3 that

$$\psi_\lambda(kxH) = \int_{K \cap H} \varphi_{\mu_\lambda}(k\ell) \exp\langle -\lambda-\rho, H(x^{-1}\ell)\rangle \, d\ell$$

for $k \in K$. Therefore

$$|\psi_\lambda(kxH)| \leq \sup_{y \in K} |\varphi_{\mu_\lambda}(y)| \cdot \psi_\lambda(xH) .$$

From Theorem 8.1.1 it then follows that it suffices to prove

$$\int_{\mathfrak{b}_o^+} [\psi_\lambda(\exp Y)]^2 \, \delta(Y) \, dY < \infty \quad . \tag{8.10}$$

Recall that $\mathfrak{b}_o^+$ is a positive chamber for $\Sigma_o = \Sigma(\mathfrak{b}, \mathfrak{g}_o)$. Then $\overline{\mathfrak{b}_o^+}$ is a union of closed chambers $\overline{\mathfrak{b}^+}$ for $\Sigma_b = \Sigma(\mathfrak{b}, \mathfrak{g})$, and to prove (8.10) it suffices to integrate over $\mathfrak{b}^+$ for an arbitrary

$\mathfrak{b}^+ \subset \mathfrak{b}_0^+$. Then we can use the estimate of Corollary 7.6.4, which ensures that

$$(8.11) \qquad \psi_\lambda(\exp Y) \leq C \exp(-(1+\varepsilon)\hat{\rho}(Y))$$

for some constants C and $\varepsilon > 0$ (depending on λ but not on Y). Here $\hat{\rho}(Y) = \frac{1}{2}\Sigma_{\alpha \in \Sigma_b^+}(p_\alpha + q_\alpha)\alpha(Y)$. By its definition, the order of growth of $\delta(Y)$ as Y tends to infinity is $\sim \exp 2\hat{\rho}(Y)$. Since $\hat{\rho}(Y) > 0$ on $\mathfrak{b}^+$, the estimate (8.11) then ensures that

$$\int_{\mathfrak{b}^+} [\psi_\lambda(\exp Y)]^2 \, \delta(Y) \, dY < \infty \quad . \quad \square$$

If G is not linear, then it easily follows from Theorem 8.2.4 that we can still define functions $\psi_{w,\lambda} \in C_K^\infty(G/H)$ provided (8.8) and (8.9) hold (where $\exp X$ in (8.9) is defined in G). Also, Theorem 8.3.1 holds as stated, for G semisimple with finite center. If G has not finite center one can also define Flensted-Jensen functions, and Theorem 8.3.1 still holds, provided "square integrable" is interpreted in the right sense, taking into consideration the non-compactness of K - see Flensted-Jensen [c] p. 269.

Under the assumption of Theorem 8.3.1 let $\pi_{w,\lambda}$ denote the subrepresentation of $L^2(G/H)$ which $\psi_{w,\lambda}$ generates. The following theorem shows that $\pi_{w,\lambda}$ is actually in the discrete series of $L^2(G/H)$.

<u>Theorem 8.3.2</u> (<u>i</u>) <u>The K-type</u> $\mu^\vee_{w,\lambda}$ <u>has multiplicity one in</u> $\pi_{w,\lambda}$.

(<u>ii</u>) <u>The representation</u> $\pi_{w,\lambda}$ <u>of</u> G <u>is irreducible</u>.

<u>Proof</u>: (i) This follows from a simple algebraic result due to Lepowsky and McCollum (cf. Dixmier [a] Proposition 9.1.10 (iii)), since $\psi_{w,\lambda}$ is cyclic, K-finite of irreducible type $\mu^\vee_{w,\lambda}$, and a joint eigenvector for $U(\mathfrak{g})^K$.

(ii) Since $\pi_{w,\lambda}$ has an infinitesimal character it is the closure of the direct sum of its irreducible subrepresentations (cf. Harish-Chandra [a] Theorem 7). Since $\psi_{w,\lambda}$ is cyclic each of these subrepresentations must contain the K-type $\mu^\vee_{w,\lambda}$ with positive multiplicity. Therefore (i) implies (ii). $\square$

For further properties of the K-types of $\pi_{w,\lambda}$, see Flensted-Jensen [c] Theorem 6.5 and Schlichtkrull [b] Section 5.

Corollary 8.3.3 *If* rank G/H = rank $K/K\cap H$ *then the discrete series for* G/H *is not empty.*

8.4 Examples

In this section we will briefly mention some examples. For further examples, see the references mentioned in the notes at the end of this chapter.

8.4.1 Discrete series for G

Let G be a connected noncompact semisimple Lie group. For simplicity we assume $G \subset G^{\mathbb{C}}$, with $G^{\mathbb{C}}$ simply connnected. Let $G^{\times} = G \times G$ and $H^{\times} = \text{diag } G$, then $G^{\times}/H^{\times}$ is a semisimple symmetric space, diffeomorphic as a manifold to G (cf. Section 7.1, Example b). It is easily seen that invariant measure on $G^{\times}/H^{\times}$ up to scalars is identical to Haar measure on G. Let $\mathfrak{g} = \mathfrak{k} \oplus \mathfrak{p}$ be a Cartan decomposition of $\mathfrak{g}$ and K the corresponding maximal compact subgroup. Then $K^{\times} = K \times K$ is maximally compact in $G^{\times}$. Let $\mathfrak{a} \subset \mathfrak{p}$ be a maximal abelian subspace, then $\mathfrak{b}^{\times} = \{(X,-X) \mid X \in \mathfrak{a}\}$ is a maximal abelian subspace of $\mathfrak{p}^{\times} \cap \mathfrak{q}^{\times}$ where $\mathfrak{p}^{\times} = \mathfrak{p} \times \mathfrak{p}$ and $\mathfrak{q}^{\times} = \{(X,-X) \mid X \in \mathfrak{g}\}$. The formula in Theorem 8.1.1 then takes the form (which is well known):

$$\int_G f(g)dg = \int_K \int_{\mathfrak{a}^+} \int_K f(k_1 \exp X\, k_2)\delta(X)dk_1 dk_2 dX$$

where $\delta(X) = \prod_{\alpha \in \Sigma^+} |\cosh \alpha(X) \sinh \alpha(X)|^{m_\alpha}$.

Let θ denote the Cartan involution of $\mathfrak{g}^{\mathbb{C}}$, that is, complex conjugation with respect to the real form $\mathfrak{k} + \sqrt{-1}\,\mathfrak{p}$. Let
$G_{\mathbb{C}} = \{(z,\theta z) \mid z \in G^{\mathbb{C}}\}$, $K_{\mathbb{C}} = \{(z,\theta z) \mid z \in K^{\mathbb{C}}\}$ and
$U = \{(u,u) \mid u \in \exp(\mathfrak{k} + \sqrt{-1}\,\mathfrak{p})\}$, then these are the dual objects: $G^{\times o} = G_{\mathbb{C}}$, $K^{\times o} = K_{\mathbb{C}}$ and $H^{\times o} = U$, since both $G^{\times}$ and $G_{\mathbb{C}}$ lie inside $G^{\mathbb{C}} \times G^{\mathbb{C}}$. Of course, $z \longrightarrow (z,\theta z)$ is an isomorphism $G^{\mathbb{C}} \xrightarrow{\sim} G_{\mathbb{C}}$.

Now Theorem 8.2.1 gives a bijection between functions on G which are both right and left K-finite, and $K_{\mathbb{C}}$-finite functions on $G_{\mathbb{C}}/U$.

Assume that $\mathfrak{g}$ has a compact Cartan subalgebra $\mathfrak{t}$, and let $\mathfrak{t}^{\times} = \{(H,-H) \mid H \in \mathfrak{t}\}$, then $\mathfrak{t}^{\times}$ is a Cartan subspace for $G^{\times}/H^{\times}$ which is contained in $\mathfrak{h}^{\times}$. Thus we have that $\operatorname{rank} G^{\times}/H^{\times} = \operatorname{rank} K^{\times}/K^{\times} \cap H^{\times}$. Let Δ denote the root system of $\mathfrak{t}$ in $\mathfrak{g}_c$, and Δ_c that of $\mathfrak{t}$ in $\mathfrak{k}_c$. Then $\Sigma = \Sigma(\mathfrak{t}^{\times}, \mathfrak{g}_{\mathbb{C}})$ consists of the roots α given by $\alpha(H,-H) = \beta(H)$ $(\beta \in \Delta)$, each with multiplicity 2, and similarly the roots of $\Sigma_c = \Sigma(\mathfrak{t}^{\times}, \mathfrak{k}_{\mathbb{C}})$ are given by the same equation $(\beta \in \Delta_c)$ with multiplicity 2. Fix positive systems $\Delta_c^+ \subset \Delta^+$, and Σ_c^+ and Σ^+ correspondingly.

For each $\lambda \in \mathfrak{t}_c^*$ we define $\lambda^{\times} \in (\mathfrak{t}^{\times})_c^*$ by $\lambda^{\times} = (H,-H) = 2\lambda(H)$ for $H \in \mathfrak{t}$. With $\rho = \frac{1}{2}\sum_{\alpha\in\Delta^+}\alpha$ and $\rho_c = \frac{1}{2}\sum_{\alpha\in\Delta_c^+}\alpha$ we have $\rho^{\times} = \frac{1}{2}\sum_{\alpha\in\Sigma^+}(\dim \mathfrak{g}_{\mathbb{C}}^{\alpha})$ and similarly for ρ_c. With $\mu_\lambda = \lambda + \rho - 2\rho_c$ we then get $\mu_\lambda^{\times} = \lambda^{\times} + \rho^{\times} - 2\rho_c^{\times}$. Notice that the condition

$$\frac{<\mu_\lambda^{\times}, \beta>}{<\beta,\beta>} \in \mathbb{Z}_+ , \text{ for all } \beta \in \Sigma_c^+$$

is equivalent to

$$\frac{2<\mu_\lambda, \alpha>}{<\alpha,\alpha>} \in \mathbb{Z}_+ , \quad \text{for all } \alpha \in \Delta_c^+$$

that is, μ_λ is the highest weight of a K-type.

Let $\xi : U(\mathfrak{g})^K \longrightarrow U(\mathfrak{t})$ denote the homomorphism uniquely determined by

$$u - \xi(u) \in U(\mathfrak{g})\,\mathfrak{n}_c$$

where $\mathfrak{n}_c = \sum_{\alpha\in\Delta^+} \mathfrak{g}_c^{\alpha}$, then the map $\xi_{\mathbb{C}} : U(\mathfrak{g}_{\mathbb{C}})^{K_{\mathbb{C}}} \longrightarrow U(\mathfrak{t}^{\times})$ defined in Proposition 7.4.1 for $G_{\mathbb{C}}/K_{\mathbb{C}}$ is given by

$$\xi_{\mathbb{C}}(u, \theta u)(\lambda^{\times}) = 2\xi(u)(\lambda)$$

for $\lambda \in \mathfrak{t}_c^*$ and $u \in U(\mathfrak{g}^{\mathbb{C}})$ (cf. the Remark following Proposition 7.4.1). As a special case of Theorem 8.3.1 we then have:

Theorem 8.4.2 (rank G = rank K) For each positive system Δ^+ compatible with Δ_c^+, and each $\lambda \in \sqrt{-1}\,\mathfrak{t}^*$ satisfying

$$<\lambda,\alpha> > 0 , \quad \forall\alpha \in \Delta^+$$

and

$$\frac{2<\mu_\lambda , \alpha>}{<\alpha , \alpha>} \in \mathbb{Z}_+ , \quad \forall\alpha \in \Delta_c^+$$

there exists a discrete series representation π_λ of G with the following properties:

(i) π_λ contains the K-type $\mu_\lambda^\vee$ with multiplicity one.

(ii) $U(\mathfrak{g})^K$ acts on the K-type $\mu_\lambda^\vee$ in π_λ via the scalar homomorphism: $u \longrightarrow \xi(u)(-\lambda-\rho)$.

One can prove that the representations π_λ thus constructed exhaust the discrete series for G , and that if rank K $\neq$ rank G there is no discrete series (Harish-Chandra [d]).

For a thorough treatment of this example using Flensted-Jensen functions we refer to Knapp [b] (see also Section 7 of Flensted-Jensen [c]).

8.4.2 The hyperboloids

Let $G = SO_o(p,q)$ and $H = SO_o(p,q-1)$ $(p \geq 1 , q \geq 2)$, then G/H is identified with the hypersurface

$$X = \{x \in \mathbb{R}^{p+q} \mid x_1^2 + \ldots + x_p^2 - x_{p+1}^2 - \ldots - x_{p+q}^2 = -1\} .$$

We take $K = SO(p) \times SO(q)$ and $\mathfrak{b} = \mathbb{R}\,Y_1$ where $Y_1 = E_{1,p+q} + E_{p+q,1}$, then $\mathfrak{b}$ is maximal abelian in $\mathfrak{p} \cap \mathfrak{q}$ (and in $\mathfrak{q}$). We have $\Sigma_b = \{\pm\beta\}$ where $\beta(Y_1) = 1$ with multiplicity $p_\beta + q_\beta = p+q-2$. Moreover $G_o = SO_o(p,1)$ and the multiplicity of β in $\mathfrak{g}_o$ is $p_\beta = p-1$. Thus $q_\beta = q-1$, and we have

$$\delta(tY_1) = |\sinh t|^{p-1}(\cosh t)^{q-1} .$$

We have $K \cap H = SO(p) \times SO(q-1)$ and the centralizer of $\mathfrak{b}$ in $K \cap H$ is $M_b = SO(p-1) \times SO(q-1)$. Let

$$Y = S^{p-1} \times S^{q-1} = \{y = (y', y'') \in \mathbb{R}^p \times \mathbb{R}^q \mid \Sigma y_i'^2 = \Sigma y_j''^2 = 1\}$$

and define $\Phi : Y \times \mathbb{R}_+ \longrightarrow X$ by

$$\Phi(y,t) = (y_1' \sinh t, \ldots, y_p' \sinh t, y_1'' \cosh t, \ldots, y_q'' \cosh t)$$

then Φ is an analytic isomorphism onto an open dense subset of X. Then Φ is the map of Section 8.1 (except in case $p = 1$, where Y is not connected and $\mathfrak{b}_0^+ = \mathfrak{b}$). From Theorem 8.1.1 we then have the integration formula

$$(8.12) \qquad \int_X f(x)dx = \int_Y \int_0^\infty f(\Phi(y,t)) \sinh^{p-1} t \cosh^{q-1} t \, dt \, dy$$

(which holds also for $p = 1$).

The dual spaces are $G^o = SO_o(p+q-1,1)$, $H^o = SO(p+q-1)$ and $K^o = SO(p) \times SO_o(q-1,1)$, and thus G^o/K^o is of the type considered in Section 7.7.1.

Let $\mathfrak{t} = \mathbb{R}X$ where $X = E_{p+q-1,p+q} - E_{p+q,p+q-1}$, and let $\lambda \in \mathfrak{t}_c^*$ be given by $\lambda(\sqrt{-1}\,X) = a \in \mathbb{C}$. Then $\mu_\lambda(\sqrt{-1}\,X) = \ell = a - 1 + \frac{p-q}{2}$. If $q > 2$ we get for each $\ell \in \mathbb{Z}_+$ that the Flensted-Jensen function is given by

$$(8.13) \qquad \psi_\lambda(\Phi(y,t)) = \varphi_\ell(y'')(\cosh t)^{-a+1-\frac{p+q}{2}}$$

where $\varphi_\ell(y'')$ is the spherical function $\phi_{\mu_\lambda + \rho_c}$ on S^{q-1}, which essentially is a polynomial. If $q = 2$ we have the same formula for each $\ell \in \mathbb{Z}$. Thus Flensted-Jensen's functions have very explicit expressions on the hyperboloids. Comparing (8.12) and (8.13) we see that if $a > 0$ then $\psi_\lambda \in L^2(G/H)$, as proved in Theorem 8.3.1. When $q = 2$ the function $\psi_{-1,\lambda}$ can be similarly computed.

When $q \leq p+4$ one can prove that the representations S_λ generated by these ψ_λ actually exhaust the discrete series for G/H . However, when $q > p+4$ there is a finite set of discrete series representations for G/H which can not be obtained from Flensted-Jensen's functions. This follows from Strichartz [a], where the Plancherel formula for the hyperboloids is explicitly determined. (See also Rossmann [a], Flensted-Jensen [c] Section 8, and Flensted-Jensen and Okamoto [a]).

8.4.3 $G/H = SO_o(p+1,1)/SO(2) \times SO_o(p-1,1)$

With $p > 3$ this symmetric space of rank 2 satisfies the equal rank condition. The dual space is $G^o/K^o = SO_o(2,p)/SO_o(2,p-1)$ which was treated in Section 7.7.2. We leave further details to the reader.

8.5 Notes and further results

The construction of L^2-functions on G/H in this chapter follows M. Flensted-Jensen [c]. Both the integration formula (Theorem 8.1.1) and the duality theorem (8.2.1) is from that paper. The duality was also considered by Flensted-Jensen in [b] (the dual symmetric space G^o/H^o was introduced in Berger [a] p. 111). In Flensted-Jensen [c] p. 273 an elementary proof is given for Theorem 8.3.1 with the extra condition on λ that it is sufficiently far from the walls of $\mathfrak{a}_+^*$. That the theorem holds as stated was conjectured in loc. cit., and proved by T. Oshima. The proof we give was kindly put at our disposal by T. Oshima. In Oshima and Matsuki [b] a more general statement is proved (see below). Theorem 8.3.2 as well as Examples 8.4.1 and 8.4.2 are also from Flensted-Jensen [c].

For the very important, special case of the group itself, the explicit decomposition of $L^2(G)$ (Plancherel formula) has been determined in the work of Harish-Chandra ([b], [d], [f] - see also the survey [e]). The discrete series is parametrized in [d]. The properties (i) and (ii) of Theorem 8.4.2 are proved in Hecht and Schmid [a], Schmid [a] and Wallach [b]. For further results and references on the discrete series for G we refer to Duflo [a]. For the symmetric space G/K the Plancherel formula was also proved by Harish-Chandra ([c] and [d]). An important contribution was the computation of the integral (6.7) by Gindikin and Karpelevič [a].

For the real hyperboloids (Example 8.4.2) the Plancherel formula has been explicitly determined by N. Limič, J. Niederle and R. Raczka [a], and R. Strichartz [a] (For special values of p and q see Gel'fand et al. [a], Shintani [a], and Molčanov [a]. See also Faraut [a] and Rossmann [a]). Other semisimple symmetric spaces where the Plancherel formula is explicitly known are the complex, quaternion and octonion hyperboloids (Matsumoto [a], Faraut [b] and Kosters [a]). See Flensted-Jensen and Okamoto [a] for an interpretation of the full discrete series of these spaces in the spirit of Flensted-Jensen [a]. See also the announcements Oshima [d] and Kengmana [a].

In [b], T. Oshima and T. Matsuki give a general description of the discrete series for semisimple symmetric spaces, using the duality of Section 8.2 and the boundary value maps constructed by T. Oshima [f]. Their results require the study of β_λ also when Assumption (A) does not hold. Among their results we mention the important converse to

Corollary 8.3.3 that if the rank of $K/K \cap H$ is not equal to that of G/H then there is no discrete series for G/H .(See also Oshima [g]).

Some further results and generalizations have also been announced in Flensted-Jensen [d] and [e]. In Matsumoto [b] Flensted-Jensen's representations are constructed in a special case. For the significance of the discrete series for G/H in the theory of unitary representations of G we refer to the author's paper [b]. In [a] the author generalizes the results of Sections 8.2 and 8.3 to vector bundles over G/H . Further results on Flensted-Jensen's representations are also given in Ólafsson [a], [b] and in Schlichtkrull [d].

Bibliography

Aomoto, K.

[a] On some double coset decompositions of complex semisimple Lie groups. J. Math. Soc. Japan 18 (1966), 1-44.

Araki, S.

[a] On root systems and an infinitesimal classification of irreducible symmetric spaces. J. Math. Osaka City Univ. 13 (1962), 1-34.

Benoist, Y.

[a] Sur l'algèbre des opérateurs différentiels invariant sur un espace symétrique nilpotent. C.R. Acad. Sci. Paris Sér. I Math. 295 (1982), 59-62.

Berger, M.

[a] Les espaces symétriques non compacts. Ann. Sci. École Norm. Sup. (3) 74 (1957), 85-177.

Berline, N. and Vergne, M.

[a] Équations de Hua et noyau de Poisson. Pp. 1-51 in Carmona and Vergne [a] (1981).

Björk, J.-E.

[a] Rings of Differential Operators. North-Holland Mathematical Library, Vol. 21, North-Holland Publishing Co., Amsterdam-New York-Oxford 1979.

Bony, J.

[a] Hyperfonctions et Équations aux Dérivées Partielles, Cours de 3e Cycle, Orsay 1974-75 (unpublished).

Bony, J. and Schapira, P.

[a] Existence et prolongement des solutions holomorphes des équations aux dérivées partielles. Invent. Math. 17 (1972), 95-105.

Boothby, W. M. and Weiss, G. L. (eds.)

[a] Symmetric Spaces. Short Courses Presented at Washington Univ., St. Louis 1969-1970. Pure and Appl. Math., Vol. 8, Marcel Dekker, New York 1972.

Bourbaki, N.

[a] Éléments de Mathématique, Groupes et Algébres de Lie. Fascicule 34, Chap. 4-6. Hermann, Paris 1968.

Boutet de Monvel, L. and Krée, P.

[a] Pseudo-differential operators and Gevrey-classes. Ann. Inst. Fourier (Grenoble) 17-1 (1967), 295-323.

Bredon, G. E.

[a] Sheaf Theory. McGraw-Hill Book Co., New York-Toronto 1967.

Bremermann, H.

[a] Distributions, Complex Variables, and Fourier Transforms. Addison-Wesley Publishing Co., Reading-London 1965.

Bruhat, F.

[a] Sur les représentations induites des groupes de Lie. Bull. Soc. Math. France 84 (1956), 97-205.

Cahen, M. and Parker, M.

[a] Pseudo-Riemannian Symmetric Spaces. Mem. Amer. Math. Soc., Vol. 24, no. 229, Amer. Math. Soc., Providence 1980.

Carmona, J. and Vergne, M. (eds.)

[a] Non Commutative Harmonic Analysis and Lie groups, Proceedings, Marseillie-Luminy 1980. Lecture Notes in Math. 880, Springer-Verlag, Berlin-New York 1981.

Cartan, E.

[a] Sur la détermination d'un système orthogonal complet dans un espace de Riemann symmetrique clos. Rend. Circ. Mat. Palermo 53 (1929), 217-252.

Casselman, W. and Miličić, D.

[a] Asymptotic behavior of matrix coefficients of admissible representations. Duke Math. J. 49 (1982), 869-930.

Cerezo, A., Chazarain, J. and Piriou, A.

[a] Introduction aux hyperfonctions. Pp. 1-53 in Pham [a] (1975).

Chang, W.

[a] Global solvability of the Laplacians on pseudo-Riemannian symmetric spaces. J. Funct. Anal. 34 (1979), 481-492.

Coddington, E. A. and Levinson, N.

[a] Theory of Ordinary Differential Equations. McGraw-Hill, New York 1955.

Cygan, J.

[a] A tangential convergence for bounded harmonic functions on a rank one symmetric space. Trans. Amer. Math. Soc. 265 (1981), 405-418.

Deligne, P.

[a] Équations Différentielles à Points Singuliers Réguliers. Lecture Notes in Math. 163, Springer-Verlag, Berlin-New York 1970.

Dixmier, J.

[a] Algèbres Enveloppantes. Cahiers Scientifiques, Vol. 37, Gauthiers-Villars Éditeur, Paris-Bruxelles-Montréal 1974.

Duflo, M.

[a] Représentations de carré intégrable des groupes semisimple réels. Sém. Bourbaki, Exp. 508, pp. 23-40. Lecture Notes in Math. 710, Springer-Verlag, Berlin-New York 1979.

Erdélyi, A. et al.

[a] Higher Transcendental Functions (Bateman Manuscript Project), Vol. 1. McGraw-Hill, New York 1953.

Faraut, J.

[a] Noyaux sphériques sur un hyperboloïde á une nappe. Pp. 172-210 in Analyse Harmonique sur les Groupes de Lie, Séminaire Nancy-Strasbourg 1973-1975, Eymard, P. et al. (eds.). Lecture Notes in Math. 497, Springer-Verlag, Berlin-New York 1975.

[b] Distributions sphériques sur les espaces hyperboliques. J. Math. Pures Appl. (9) 58 (1979), 369-444.

Flensted-Jensen, M.

[a] Spherical functions on rank one symmetric spaces and generalizations. Pp. 339-342 in Moore [b] (1973).

[b] Spherical functions on a real semisimple Lie group. A method of reduction to the complex case. J. Funct. Anal. 30 (1978), 106-146.

[c] Discrete series for semisimple symmetric spaces. Ann. of Math. (2) 111 (1980), 253-311.

[d] K-finite joint eigenfunctions of $U(g)^K$ on a non-Riemannian semisimple symmetric space G/H . Pp. 91-101 in Carmona and Vergne [a] (1981).

[e] Harmonic analysis on semisimple symmetric spaces. A method of duality. To appear in Herb et al. [a] (1984).

Flensted-Jensen, M. and Okamoto, K.

[a] An explicit construction of the K-finite vectors in the discrete series for an isotropic semisimple symmetric space. To appear in the proceedings of the conference in Kleebach, France 1983, Mém. Soc. Math. France (N.S.).

Furstenberg, H.

[a] A Poisson formula for semisimple Lie groups. Ann. of Math. (2) 77 (1963), 335-386.

[b] Translation-invariant cones of functions on semisimple Lie groups. Bull. Amer. Math. Soc. 71 (1965), 271-326.

Gel'fand, I. M., Graev, M. I., and Vilenkin, N. Y.

[a] Generalized Functions, Vol. 5: Integral Geometry and Representation Theory. Academic Press, New York-London 1966.

Gindikin, S. G. and Karpelevič, F. I.

[a] Plancherel measure for Riemann symmetric spaces of non-positive curvature. Sov. Math. 3 (1962), 962-965 (transl. from Dokl. Akad. Nauk SSSR 145 (1962), 252-255).

Grauert, H.

[a] On Levi's problem and the imbedding of real-analytic manifolds. Ann. of Math. (2) 68 (1958), 460-472.

Grauert, H. and Remmert, R.

[a] Theory of Stein Spaces. Die Grundlehren der Math. Wiss., Vol.236, Springer-Verlag, Berlin-New York 1979.

Grothendieck, A.

[a] Local Cohomology. Notes by R. Hartshorne. Lecture Notes in Math. 41, Springer-Verlag, Berlin-New York 1967.

Harish-Chandra

[a] Representations of a semisimple Lie group on a Banach space, I. Trans. Amer. Math. Soc. 75 (1953), 185-243.

[b] The Plancherel formula for complex semisimple Lie groups. Trans. Amer. Math. Soc. 76 (1954), 485-528.

[c] Spherical functions on a semisimple Lie group,

I. Amer. J. Math. 80 (1958), 241-310.

II. Ibid, pp. 553-613.

[d] Discrete series for semisimple Lie groups,

I. Acta Math. 113 (1965), 241-318.

II. Acta Math. 116 (1966), 1-111.

[e] Harmonic analysis on semisimple Lie groups. Bull. Amer. Math. Soc. 76 (1970), 529-551.

[f] Harmonic analysis on real reductive groups,

I. The theory of the constant term. J. Funct. Anal. 19 (1975), 104-204.

II. Wave packets in the Schwartz space. Invent. Math. 36 (1976), 1-55.

III. The Maass-Selberg relations and the Plancherel formula. Ann. of Math. (2) 104 (1976), 117-201.

[g] Some results on differential equations. (Unpublished 1960). Pp. 7-56 in Collected Papers, Vol. 3, ed. Varadarajan, V. S., Springer-Verlag, Berlin-New York 1984.

Hartshorne, R.

[a] Residues and Duality. Lecture Notes in Math. 20, Springer-Verlag, Berlin-New York 1966.

Hashizume et al.

[a] Hashizume, M., Kowata, A., Minemura, M., and Okamoto, K. An integral representation of an eigenfunction of the Laplacian on the Euclidean space. Hiroshima Math. J. 2 (1972), 535-545.

[b] Hashizume, M., Minemura, M., and Okamoto, K., Harmonic functions on Hermitian hyperbolic spaces. Hiroshima Math. J. 3 (1973), 81-108.

Hecht, H. and Schmid, W.

[a] A proof of Blattner's conjecture. Invent. Math. 31 (1975), 129-154.

Helgason, S.

[a] Some remarks on the exponential mapping for an affine connection. Math. Scand. 9 (1961), 129-146.

[b] Differential Geometry and Symmetric Spaces. Pure and Appl. Math., Vol. 12, Academic Press, New York 1962.

[c] A duality for symmetric spaces with applications to group representations. Adv. in Math. 5 (1970), 1-154.

[d] Harmonic analysis in the non-Euclidean disk. Pp. 151-156 in Conference on Harmonic Analysis, College Park, Maryland 1971, Gulick, D. et al. (eds.). Lecture Notes in Math. 266, Springer-Verlag, Berlin-New York 1972.

[e] The surjectivity of invariant differential operators on symmetric spaces, I. Ann. of Math. (2) 98 (1973), 451-479.

[f] Eigenspaces of the Laplacian; integral representations and irreducibility. J. Funct. Anal. 17 (1974), 328-353.

[g] A duality for symmetric spaces with applications to group representations, II. Differential equations and eigenspace representations. Adv. in Math. 22 (1976), 187-219.

[h] Invariant differential equations on homogeneous manifolds. Bull. Amer. Math. Soc. 83 (1977), 751-774.

[i] Some results on eigenfunctions on symmetric spaces and eigenspace representations. Math. Scand. 41 (1977), 79-89.

[j] Differential Geometry, Lie Groups and Symmetric Spaces. Pure and Appl. Math., Vol. 80, Academic Press, New York-London 1978.

[k] Invariant differential operators and eigenspace representations. Pp. 236-286 in Representation Theory of Lie Groups, Proceedings of the SRC/LMS Research Symposium Oxford 1977, Atiyah, M.F.(ed.). London Math. Soc. Lecture Notes, Vol. 34, Cambridge University Press, Cambridge 1979.

[l] A duality for symmetric spaces with applications to group representations, III. Tangent space analysis. Adv. in Math. 36 (1980), 297-323.

[m] Topics in Harmonic Analysis on Homogeneous Spaces. Progress in Math., Vol. 13, Birkhäuser, Boston-Basel-Stuttgart 1981.

[n] Groups and Geometric Analysis. Integral Geometry, Invariant Differential Operators and Spherical Functions. Academic Press, New York (to appear).

Helgason, S. and Johnson, K. D.

[a] The bounded spherical functions on symmetric spaces. Adv. in Math. 3 (1969), 583-593.

Helgason, S. and Korányi, A.

[a] A Fatou-type theorem for harmonic functions on symmetric spaces. Bull. Amer. Math. Soc. 74 (1968), 258-263.

Herb, R. et al. (eds.)

[a] Lie Group Representations, Proceedings Maryland 1982-1983,

I. Lecture Notes in Math. 1024, Springer-Verlag, Berlin-New York 1983.

II. Lecture Notes in Math. 1041, Springer-Verlag, Berlin-New York 1984.

III. Lecture Notes in Math. , Springer-Verlag, Berlin-New York 1984 (to appear).

Hilb, E.

[a] Lineare Differentialgleichungen im komplexen Gebiet. Encyklopädie der mathematischen Wissenschaften II B 5 (1915).

Hilton, P. J. and Stammbach, U.

[a] A Course in Homological Algebra. Graduate Texts in Math., Vol. 4, Springer-Verlag, Berlin-New York 1971.

Hiraoka, K., Matsumoto, S., and Okamoto, K.

[a] Eigenfunctions of the Laplacian on a real hyperboloid of one sheet. Hiroshima Math. J. 7 (1977), 855-864.

Hirzebruch, F.

[a] Topological Methods in Algebraic Geometry. Die Grundlehren der Math. Wiss., Vol. 131, Springer-Verlag, New York 1966.

Hoogenboom, B.

[a] The generalized Cartan decomposition for a compact Lie group. Preprint, Amsterdam 1983.

[b] Intertwining functions on compact Lie groups. Thesis, Univ. of Leiden 1983.

Hörmander, L.

[a] An Introduction to Complex Analysis in Several Variables. D. van Nostrand Co., Princeton-Toronto-London 1966.

[b] The Analysis of Linear Partial Differential Operators, I. Springer-Verlag, Berlin-New York 1983.

Hua, L. K.

[a] Harmonic Analysis of Functions of Several Complex Variables in the Classical Domains. Trans. of Math. Monographs, Vol. 6, Amer. Math. Soc., Providence 1963.

Inoue, T., Okamoto, K., and Tanaka, M.

[a] An integral representation of an eigenfunction of invariant differential operators on a symmetric space. Hiroshima Math. J. 4 (1974), 413-419.

Johnson, K. D.

[a] Remarks on a theorem of Korányi and Malliavin on the Siegel upper half plane of rank two. Proc. Amer. Math. Soc. 67 (1977), 351-356.

[b] Differential equations and the Bergman-Šilov boundary on the Siegel upper half plane. Ark. Mat. 16 (1978), 95-108.

[c] Generalized Hua-operators and parabolic subgroups. The cases of $SL(n,\mathbb{C})$ and $SL(n,\mathbb{R})$. Trans. Amer. Math. Soc. 281 (1984), 417-429.

[d] Generalized Hua-operators and parabolic subgroups. Preprint.

Johnson K. D. and Korányi, A.

[a] The Hua operators on Hermitian symmetric domains of tube type. Ann. of Math. (2) 111 (1980), 589-608.

Jørgensen, S. T.

[a] Hyperfunktioner og deres anvendelser i differentialligningsteori (in Danish). Speciale, Københavns Universitets Matematiske Institut 1976.

Karpelevič, F. I.

[a] The simple subalgebras of the real Lie algebras (in Russian). Trudy Moscov. Math. Obshch. 4 (1955), 3-112.

[b] The geometry of geodesics and the eigenfunctions of the Laplace-Beltrami operator on symmetric spaces. Trans. Moscow Math. Soc., Vol. 14, 1965, 51-199 (transl. from Trudy Moscov. Math. Obshch. 14 (1965), 48-185).

Kashiwara, M.

[a] Introduction to the theory of hyperfunctions. Pp. 3-38 in Seminar on Micro-Local Analysis, Guillemin, V., Kashiwara, M. and Kawai, T. Ann. of Math. Stud. 93, Princeton University Press, Princeton 1979.

[b] Systems of Microdifferential Equations. Progress in Math., Vol. 34, Birkhäuser, Boston-Basel-Stuttgart 1983.

Kashiwara, M. and Kawai, T.

[a] Pseudodifferential operators in the theory of hyperfunctions. Proc. Japan Acad. 46 (1970), 1130-1134.

Kashiwara et al.

[a] Kashiwara, M., Kowata, A., Minemura, K., Okamoto, K., Oshima, T., and Tanaka, M., Eigenfunctions of invariant differential operators on a symmetric space. Ann. of Math. (2) 107 (1978), 1-39.

Kashiwara, M. and Oshima, T.

[a] Systems of differential equations with regular singularities and their boundary value problems. Ann. of Math. (2) 106 (1977), 145-200.

Kawai, T.

[a] Pseudo-differential operators acting on the sheaf of microfunctions. Pp. 54-69 in Pham [a] (1975).

Kengmana, T.

[a] Characters of the discrete series for pseudo-Riemannian symmetric spaces. Pp. 177-183 in Trombi [a] (1983).

Knapp, A.W.

[a] Fatou's theorem for symmetric spaces: I. Ann. of Math. (2) 88 (1968), 106-127.

[b] Representation Theory of Semisimple Groups: An Overview Based on Examples. Forthcoming book.

Knapp, A. W. and Williamson, R. E.

[a] Poisson integrals and semisimple groups. J. Analyse Math. 24 (1971), 53-76.

Koh, S. S.

[a] On affine symmetric spaces. Trans. Amer. Math. Soc. 119 (1965), 291-309.

Komatsu, H.

[a] Boundary values for solutions of elliptic equations. Pp. 107-121 in Tokyo [a] (1970).

[b] (ed.) Hyperfunctions and Pseudo-Differential Equations, proceedings of a conference at Katata 1971. Lecture Notes in Math 287, Springer-Verlag, Berlin-New York 1973.

[c] An introduction to the theory of hyperfunctions. Pp. 3-40 in Komatsu [b] (1973).

[d] Relative cohomology of sheaves of solutions of differential equations. Pp. 192-261 in Komatsu [b] (1973).

Korányi, A.

[a] Boundary behavior of Poisson integrals on symmetric spaces. Trans. Amer. Math. Soc. 140 (1969), 393-409.

[b] Generalizations of Fatou's theorem to symmetric spaces. Rice Univ. Studies 56 (1970), 127-136.

[c] Harmonic functions on symmetric spaces. Pp. 379-412 in Boothby and Weiss [a] (1972).

[d] Poisson integrals and boundary components of symmetric spaces. Invent. Math. 34 (1976), 19-35.

[e] Compactifications of symmetric spaces and harmonic functions. Pp. 341-366 in Analyse Harmonique sur les Groupes de Lie II, Séminaire Nancy-Strasbourg 1976-78, Eymard, P. et al. (eds.). Lecture Notes in Math. 739, Springer-Verlag, Berlin-New York 1979.

[f] A survey of harmonic functions on symmetric spaces. Pp. 323-344 in Harmonic Analysis in Euclidean Spaces, Symposium in Pure Mathematics, Williams College 1978, Weiss, G. et al. (eds.). Proc. Symp. Pure Math. 35, Amer. Math. Soc., Providence 1979.

Korányi, A. and Malliavin, P.

[a] Poisson formula and compound diffusion associated to an overdetermined elliptic system on the Siegel halfplane of rank two. Acta Math. 134 (1975), 185-209.

Korányi, A. and Putz, R. B.

[a] Local Fatou theorem and area theorem for symmetric spaces of rank one. Trans. Amer. Math. Soc. 224 (1976), 157-168.

Korányi, A. and Taylor, J. C.

[a] Fine convergence and admissible convergence for symmetric spaces of rank one. Trans. Amer. Math. Soc. 263 (1981), 169-181.

Kostant, B.

[a] On convexity, the Weyl group and the Iwasawa decomposition. Ann. Sci. École Norm. Sup. (4) 6 (1973), 413-455.

Kostant, B. and Rallis, S.

[a] Orbits and representations associated with symmetric spaces. Amer. J. Math. 93 (1971), 753-809.

Kosters, M. T.

[a] Spherical distributions on rank one symmetric spaces. Thesis, Univ. of Leiden 1983.

Kowata, A. and Okamoto, K.

[a] Harmonic functions and the Borel-Weil theorem. Hiroshima Math. J. 4 (1974), 89-97.

Kowata, A. and Tanaka, M.

[a] Global solvability of the Laplace operator on a noncompact affine symmetric space. Hiroshima Math. J. 10 (1980), 409-417.

Lassalle, M.

[a] Transformées de Poisson, algèbres de Jordan et équations de Hua. C. R. Acad. Sci. Paris Sér. I Math. 294 (1982), 325-328.

[b] Algèbres de Jordan, coordonnées polaires et équations de Hua. C. R. Acad. Sci. Paris Sér. I Math. 294 (1982), 613-615.

Lewis, J. B.

[a] Eigenfunctions on symmetric spaces with distribution valued boundary forms. J. Funct. Anal. 29 (1978), 287-307.

Limič, N., Niederle, J., and Rączka, R.

[a] Eigenfunction expansions associated with second-order invariant operator on hyperboloids and cones, III. J. Math. Phys. 8 (1967), 1079-1093.

Lindahl, L.-Å.

[a] Fatou's theorem for symmetric spaces. Ark. Mat. 10 (1972), 33-47.

Loos, O.

[a] Symmetric Spaces, I: General Theory. W. A. Benjamin Inc., New York-Amsterdam 1969.

Lowdenslager, D. B.

[a] Potential theory in bounded symmetric homogeneous complex domains. Ann. of Math. 67 (2) (1958), 467-484.

Lützen, J.

[a] The Prehistory of the Theory of Distributions. Springer-Verlag, Berlin-New York 1982.

Malgrange, B.

[a] Faisceaux sur des variétés analytiques réelles. Bull. Soc. Math. France 85 (1957), 231-237.

Mantero, A. M.

[a] Modified Poisson kernels on rank one symmetric spaces. Proc. Amer. Math. Soc. 77 (1979), 211-217.

Martineau, A.

[a] Les hyperfonctions de M. Sato. Séminaire Bourbaki, Volume 1960/1961 (13e année), n° 214, Benjamin, New York 1966.

[b] Le "edge of the wedge theorem" en théorie des hyperfonctions de Sato. Pp. 95-106 in Tokyo [a] (1970).

Matsuki, T.

[a] The orbits of affine symmetric spaces under the action of minimal parabolic subgroups. J. Math. Soc. Japan 31 (1979), 331-357.

[b] Orbits on affine symmetric spaces under the action of parabolic subgroups. Hiroshima Math. J. 12 (1982), 307-320.

Matsumoto, S.

[a] The Plancherel formula for a pseudo-Riemannian symmetric space. Hiroshima Math. J. 8 (1978), 181-193.

[b] Discrete series for an affine symmetric space. Hiroshima Math. J. 11 (1981), 53-79.

Michelson, H. L.

[a] Fatou theorems for eigenfunctions of the invariant differential operators on symmetric spaces. Trans. Amer. Math. Soc. 177 (1973), 257-274.

[b] Generalized Poisson integrals and their boundary behavior. Pp. 329-333 in Moore [b] (1973).

Minemura, K.

[a] Harmonic functions on real hyperbolic spaces. Hiroshima Math. J. 3 (1973), 121-151.

[b] Eigenfunctions of the Laplacian on a hermitian hyperbolic space. Hiroshima Math. J. 4 (1974), 441-457.

[c] Eigenfunctions of the Laplacian on a real hyperbolic space. J. Math. Soc. Japan 27 (1975), 82-105.

Miwa, T., Oshima, T., and Jimbo, M.

[a] Introduction to microlocal analysis. Publ. Res. Inst. Math. Sci. 12 Suppl. (1977), 267-300.

Molčanov, V. F.

[a] Analogue of the Plancherel formula for hyperboloids. Soviet Math. Dokl. 9 (1968), 1382-1385 (transl. from Dokl. Akad. Nauk SSSR 183 (1968), 288-291).

Moore, C. C.

[a] Compactifications of symmetric spaces,

I. Amer. J. Math. 86 (1964), 201-218.

II. Ibid., pp. 358-378.

[b] (ed.) Harmonic Analysis on Homogeneous Spaces, Symposium in Pure Mathematics, Williams College 1972. Proc. Symp. Pure Math. 26, Amer. Math. Soc., Providence 1973.

Morimoto, M.

[a] Edge of the wedge theorem and hyperfunction. Pp. 41-81 in Komatsu [b] (1973).

[b] A generalization of the Fourier-Borel transformation for the analytic functionals with non convex carrier. Tokyo J. Math. 2 (1979), 301-321.

Mostow, G. D.

[a] Some new decomposition theorems for semi-simple groups. Mem. Amer. Math. Soc. 14 (1955), 31-54.

[b] On covariant fiberings of Klein spaces. Amer. J. Math. 77 (1955), 247-278.

Nomizu, K.

[a] Invariant affine connections on homogeneous spaces. Amer. J. Math. 76 (1954), 33-65.

Ólafsson, G.

[a] Die Langlands-Parameter für die Flensted-Jensensche fundamentale Reihe. Preprint Univ. of Iceland 1983.

[b] Die Darstellungsreihe zu einem affinen symmetrischen Raum. Preprint Univ. of Iceland 1983.

Oshima, T.

[a] A realization of Riemannian symmetric spaces. J. Math. Soc. Japan 30 (1978), 117-132.

[b] Poisson transformation of affine symmetric spaces. Proc. Jap. Acad. Ser. A Math. Sci. 55 (1979), 323-327.

[c] Communication with Flensted-Jensen (1980).

[d] Fourier analysis on semisimple symmetric spaces. Pp. 357-369 in Carmona and Vergne [a] (1981).

[e] A definition of boundary values of solutions of partial differential equations with regular singularities. Publ. Res. Inst. Math. Sci. 19 (1983), 1203-1230.

[f] Boundary value problems for systems of linear partial differential equations with regular singularities. Preprint, 1983.

[g] Discrete series for semisimple symmetric spaces. Proceedings of the International Congress of Mathematicians Warsaw 1983 (to appear).

Oshima, T. and Matsuki, T.

[a] Orbits on affine symmetric spaces under the action of the isotropy subgroups. J. Math. Soc. Japan 32 (1980), 399-414.

[b] A description of discrete series for semisimple symmetric spaces. Preprint 1983.

Oshima, T. and Sekiguchi, J.

[a] Eigenspaces of invariant differential operators on an affine symmetric space. Invent. Math. 57 (1980), 1-81.

Pham, F. (ed.)

[a] Hyperfunctions and Theoretical Physics, Rencontre de Nice 1973. Lecture Notes in Math. 449, Springer-Verlag, Berlin-New York 1975.

Rossmann, W.

[a] Analysis on real hyperbolic spaces. J. Funct. Anal. 30 (1978), 448-477.

[b] The structure of semisimple symmetric spaces. Canad. J. Math. 31 (1979), 157-180.

Rudin, W.

[a] Real and Complex Analysis. McGraw-Hill Book Co., New York-Düsseldorf-Johannesburg 1974.

Satake, I.

[a] On representations and compactifications of symmetric Riemannian spaces. Ann. of Math. (2) 71 (1960), 77-110.

Sato, M.

[a] Theory of hyperfunctions,

I. J. Fac. Sci. Univ. Tokyo Sect. I 8 (1959-60), 139-193.

II. Ibid, pp. 387-437.

[b] Hyperfunctions and partial differential equations. Pp. 91-94 in Tokyo [a] (1970).

[c] Regularity of hyperfunction solutions of partial differential equations. Actes, Congres Intern. Math. Nice 1970, Tome 2, 785-794.

Sato, M., Kawai, T., and Kashiwara, M.

[a] Microfunctions and pseudo-differential equations. Pp. 265-529 in Komatsu [b] (1973).

Schapira, P.

[a] Théorie des Hyperfonctions. Lecture Notes in Math. 126, Springer-Verlag, Berlin-New York 1970.

[b] Hyperfonctions et problèmes aux limites elliptiques. Bull. Soc. Math. France 99 (1971), 113-141.

[c] Théorème d'unicité de Holmgren et opérateurs hyperboliques dans l'espace des hyperfonctions. An. Acad. Brasil. Ciênc. 43 (1971), 39-44.

Schlichtkrull, H.

[a] A series of unitary irreducible representations induced from a symmetric subgroup of a semisimple Lie group. Invent. Math. 68 (1982), 497-516.

[b] The Langlands parameters of Flensted-Jensen's discrete series for semisimple symmetric spaces. J. Funct. Anal. 50 (1983), 133-150.

[c] Applications of hyperfunction theory to representations of semisimple Lie groups. Prisopgave, Københavns Universitet 1983.

[d] On some series of representations related to symmetric spaces. To appear in the proceedings of the conference in Kleebach, France 1983, Mém. Soc. Math. France (N.S.).

[e] One dimensional K-types in finite dimensional representations of semisimple Lie groups. A generalization of Helgason's theorem. Math. Scand. (to appear).

[f] On the boundary behavior of generalized Poisson integrals on symmetric spaces. Preprint 1984.

Schmid, W.

[a] Some properties of square-integrable representations of semi-simple Lie groups. Ann. of Math. (2) 102 (1975), 535-564.

Sekiguchi, J.

[a] Eigenspaces of the Laplace-Beltrami operator on a hyperboloid. Nagoya Math. J. 79 (1980), 151-185.

Shafarewich, I.

[a] Basic Algebraic Geometry. Springer Study Edition, Springer-Verlag, Berlin-New York 1977.

Shapiro, R. A.

[a] Pseudo-Hermitian symmetric spaces. Comment. Math. Helv. 46 (1971), 529-548.

Shintani, T.

[a] On the decomposition of the regular representation of the Lorentz group on a hyperboloid of one sheet. Proc. Japan Acad. 43 (1967), 1-5.

Sjögren, P.

[a] Characterizations of Poisson integrals on symmetric spaces. Math. Scand. 49 (1981), 229-249.

[b] Fatou theorems and maximal functions for eigenfunctions of the Laplace-Beltrami operator in a bidisk. J. Reine Angew. Math. 345 (1983), 93-110.

[c] A Fatou theorem for eigenfunctions of the Laplace-Beltrami operator in a symmetric space. Duke Math. J. (to appear).

Stein, E. M.

[a] Boundary behavior of harmonic functions on symmetric spaces: Maximal estimates for Poisson integrals. Invent. Math. 74 (1983), 63-83.

Strichartz, R. S.

[a] Harmonic analysis on hyperboloids. J. Funct. Anal. 12 (1973), 341-383.

Sugiura, M.

[a] Representations of compact groups realized by spherical functions on symmetric spaces. Proc. Japan Acad. 38 (1962), 111-113.

Tokyo

[a] International Conference on Functional Analysis and Related Topics 1969. Proceedings, Univ. of Tokyo Press, Tokyo 1970.

Trombi, P. C. (ed.)

[a] Representation Theory of Reductive Groups: Proceedings of the University of Utah conference 1982. Progress in Math., Vol. 40, Birkhäuser, Boston-Basel-Stuttgart 1983.

Trombi, P. C. and Varadarajan, V. S.

[a] Spherical transforms on semi-simple Lie groups. Ann. of Math. (2) 94 (1971), 246-303.

Urakawa, H.

[a] Radial convergence of Poisson integrals on symmetric bounded domains of tube type. Osaka J. Math. 10 (1973), 93-113.

[b] On the Hardy class of harmonic sections and vector-valued Poisson integrals. Osaka J. Math. 12 (1975), 117-137.

van den Ban, E. P.

[a] Asymptotic expansions and integral formulas for eigenfunctions on a semisimple Lie group. Thesis, Rijksuniversiteit Utrecht 1982.

van Dijk, G.

[a] Orbits on real affine symmetric spaces. Part I: the infinitesimal case. Preprint 1982.

Varadarajan, V. S.

[a] Lie Groups, Lie Algebras, and their Representations. Prentice Hall, Englewood Cliffs, New Jersey 1974.

[b] Harmonic Analysis on Real Reductive Groups. Lecture Notes in Math. 576, Springer-Verlag, Berlin-New York 1977.

Vogan, D.

[a] Representations of Real Reductive Lie Groups. Progress in Math., Vol. 15, Birkhäuser, Boston-Basel-Stuttgart 1981.

Wallach, N.

[a] Harmonic Analysis on Homogeneous Spaces. Pure and Appl. Math., Vol. 19, Marcel Dekker Inc., New York 1973.

[b] On the Enright-Varadarajan modules: A construction of the discrete series. Ann. Sci. École Norm. Sup. (4) 9 (1976), 81-102.

[c] Asymptotic expansions of generalized matrix entries of representations of real reductive groups. Pp. 287-369 in Herb et al. [a] I (1983).

Warner, G.

[a] Harmonic Analysis on Semi-simple Lie Groups I-II. Die Grundlehren der Math. Wiss. 188-189, Springer-Verlag, Berlin-New York 1972.

Wasow, W.

[a] Asymptotic expansions for ordinary differential equations. Pure and Appl. Math., Vol. 14, Interscience Publishers, New York-London-Sydney 1965.

Weiss, N. J.

[a] Fatou's theorem for symmetric spaces. Pp. 413-441 in Boothby and Weiss [a] (1972).

Wolf, J. A.

[a] Spaces of Constant Curvature. McGraw-Hill Book Co., New York-London-Sydney 1967.

[b] The action of a real semisimple Lie group on a compact flag manifold I: Orbit structure and holomorphic arc components. Bull. Amer. Math. Soc. 75 (1969), 1121-1237.

[c] Finiteness of orbit structure for real flag manifolds. Geom. Dedicata 3 (1974), 377-384.

Zuckerman, G.

[a] Geometric methods in representation theory. Pp. 283-289 in Trombi [a] (1983).

Index